SELECTIONS FROM

SCIENCE AND SANITY

An Introduction to Non-Aristotelian Systems and General Semantics

Second Edition

Alfred Korzybski

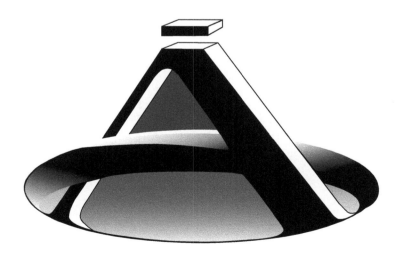

The New Non-Aristotelian Library

Institute of General Semantics   Fort Worth, Texas

© 2010 Institute of General Semantics
3000 A Landers Street, Fort Worth, Texas 76107
http://www.generalsemantics.org
First Edition 1948

Book design: Peter Darnell/Visible Works Design

ISBN: 978-0-9827559-0-7 (hardbound)
ISBN: 978-0-9827559-1-4 (paperbound)

1. General Semantics. 2. Applied Psychology. 3. Practical Philosophy.
4. Behavioral Science. 5. Communication. I. Korzybski,
Alfred. II. Kodish, Bruce, I. III. Pula, Robert P. IV. Anton, Corey.
V. Strate, Lance. VI. Title

The New Non-Aristotelian Library

Contents

Editor's Note

As Executive Director of the Institute of General Semantics, one of my priorities has been to restart the publishing program that is a vital part of the legacy of both the IGS, and the International Society for General Semantics with which it merged in 2003. And I can think of no more fitting way to begin than to bring back into circulation Alfred Korzybski's *Selections from Science and Sanity.* Originally published in 1948 for limited circulation, the book had gone through numerous reprintings in subsequent decades, and was available for a time in the form of a CD-ROM. Sadly, though, this work had been out of print for many years, depriving students of general semantics of an accessible way to encounter Korzybski's otherwise intimidating magnum opus. *Science and Sanity* is, for all its brilliance, a difficult work, due in large part to its length—927 pages! And if Korzybski himself recognized a need for an abridged edition of his introduction to general semantics and non-Aristotelian systems back in 1948, how much more so is it needed over 60 years later, in this age of text messages, status updates, and Twitter? And how much more do we need Korzybski's consciousness of abstracting at a time when so much of our lives are spent absorbed in the highly abstracted and mediated maps rendered by our digital technologies, new media, and online communications? It is true that numerous popularizations of Korzybski's general semantics have been produced over the years, some better than others, but simply put there is no substitute for encountering Korzybski himself, for reading Korzybski in his own words. Moreover, with the publication of this work, we have now brought the entire Korzybski corpus back into print, and the reader is encouraged to move from *Selections from Science and Sanity* to the unabridged work, and also to read Korzybski's first book, *Manhood of Humanity*, as well as the transcription of notes from lectures on general semantics that Korzybski gave at Olivet College published in book form, *General Semantics Seminar 1937*, and his *Collected Writings*, 1920-1950, all of which are available via the IGS. Easy access to his work, along with the publication in the near future of his biography on the part of Bruce I. Kodish, who kindly agreed to provide a Foreword for this new edition of *Selections*, should provide significant impetus for the nascent Korzybski Revival that is presently underway.

In addition to the Foreword by Bruce Kodish, this edition includes Selections from the Preface to the Fifth Edition of *Science and Sanity* by Robert P. Pula, which constitutes a wonderfully concise summary of Korzybski's thought. The main part of the text is faithful to earlier editions, with the exception of minor copyediting corrections, and a new, expanded index has been prepared, drawing on the original produced by D. David Bourland, Jr., and covering the Supplement (Colloidal Behavior) which was added in 1972 and not included in the index, as well as the new Foreword, Preface, and editorial notes. On behalf of the Institute of General Semantics, I would like to express my gratitude to Ariela Astion for copyediing and proofreading to Peter Darnell and Visible Works Design for their work on the layout and design of the cover and interior of this edition, to Judith Clarke from the Fort Worth office for her aid and assistance, to Corey Anton for serving as supervisory editor for the New Non-Aristotelian Library book series, and to Bruce Kodish for advice on all things Korzybski.

As we look towards the future of general semantics and the Institute, it is vital that we work together as time-binders, maintaining our rich heritage by bringing classic works back into print, maintaining our existing publications, and especially by making available new works that advance our discipline, to move forward along the trail that Korzybski first blazed.

Lance Strate
August 23, 2010

Series Editor's Note

This reprinting of Alfred Korzybski's *Selections from Science and Sanity* inaugurates the New Non-Aristotelian Library and represents the growing and re-enlivened dialogue regarding general semantics. Books within the series will explore the continued vitality of general semantics today.

On a recent visit to the Alfred Korzybski archive in Texas, I took a keen interest in the correspondence between Korzybski and Robert Maynard Hutchins. Both men sought to educate everyday people for the betterment of humanity but, as made evident to me in their exchange, Korzybski admired Hutchins for his efforts at mass liberal education though seemed a bit disappointed by the orientation and strategy that Hutchins had taken. Whereas Hutchins, with much assistance from Mortimer Adler, compiled and circulated the "Great Books of the Western World," Korzybski appreciated the importance of broad liberal education (and demonstrated this appreciation by his own breadth of scholarship) but he adamantly maintained the need for a teachable system, one that could help people become more conscious of abstracting and more aware of their own semantic reactions. This idea, the need for a teachable system for a more sane and mature world, continues to echo and grow stronger through time.

Great minds are often said to be slightly ahead of their time, and Korzybski proved no exception to this rule. He pioneered in the field of neurolinguistics well before the rise of contemporary neuroscience, and his work anticipated the rise of scientific culture writ large. Much like a wine that ages well, his highly prescient work grows more and more palatable as the years pass.

Anyone familiar with Korzybski's work and with the broad currents of communication theorizing in the last couple of decades of the twentieth century can be shocked at how many basic ideas come from Korzybski without due acknowledgement or adequate recognition. So many of his ideas have become staples of thought: notions such as "the map is not territory," "whatever I say a thing is, it is not," "semantic reaction," "consciousness of abstracting," and "time-binding."

Korzybski, a great synthesizer, developed his original system by drawing upon many great minds across centuries and from numerous fields. His work carries interdisciplinary, multidisciplinary, and even anti-disciplinary sensibilities as his contributions to human welfare come from insights that demand non-elementalistic orientations. In many ways, Korzybski preceded the current over-disciplinization of the academy and of knowledge, and his work harkens back to a more general education, the kind of education that seems more and more in demand given the complex and rapidly changing 21st century.

We live today in a hurried world, a world where more and more people claim to have less and less time for study. If some people feel intimidated by the sheer size of *Science and Sanity*, this authorized abridgment offers a distillation of some of Korzybski's main insights but it also hopefully serves as an introduction to the full text and encourages further study "and" practice of general semantics.

Corey Anton
August 23, 2010

Foreword to the Second Edition of
Selections from Science and Sanity

Korzybski joked that the unabridged version of *Science and Sanity*—around 800 pages long—needed its strong, sewn cloth binding in order to survive having exasperated readers pitch their copies from high-story windows. Since its 1933 publication, the book had gotten a nickname: "the blue peril." Perhaps some perils still remain within the pages of *Selections from Science and Sanity*, Korzybski's authorized abridgement of his 'magnum opus' that you now hold in your hands. But at least size should no longer be a factor of intimidation. The book was first published in the spring of 1948 in a limited print run of 3000 copies of a 274-page volume with a plain, blue, thick paper cover. It has survived pretty well over its rather long career. As far as I know, no copies ever got thrown from any great heights. Over the years and eight subsequent printings, additional material was added: an index produced by a teen-aged David Bourland in the summer of 1948, and supplemental text included several decades later. Now after having been out of print for a while, it seems like due time for this Second Edition.

When the First Edition of *Selections* was published, there existed a remarkable amount of public interest in general semantics (GS). Over the course of the decade of the 1940s and despite the war going on, GS study groups and courses had proliferated; numerous books and articles—both pro and con—had been written. There actually existed a demand for Korzybski's books. The unavailability of either of them at the start of 1948 added to the demand and the pressure that Korzybski and his colleagues at the Institute of General Semantics felt to get them out again as soon as possible. The delayed appearance of the Second Edition of Korzybski's first book, *Manhood of Humanity*, promised for several years, seemed sure to frustrate the waiting list of people who had already ordered it. (It definitely frustrated Korzybski, beginning his struggle to write a new introduction, since Manhood's theme of time-binding had once again begun to play a major role in his explicit formulating.) *Science and Sanity*, out of print now for nearly a year, had around 1000 back orders. (Korzybski was finishing an Introduction for a new Third Edition, which would be published in May.) People, especially university instructors, were eager for *Selections from Science and Sanity* as well. Publishing these last two books as soon as possible was at the top of Korzybski's to-do list.

As he wrote his "Author's Note" for *Selections*, his attention seemed especially focused on confusions about his work promulgated by some of his students, in particular S. I. Hayakawa—the well-known author and editor of *ETC*, the journal of the Society for General Semantics—whose continuing insistence on talking and writing about 'general semantics' as a kind of 'semantics', the linguistic study of 'meaning.' had by this time become very problematic from Korzybski's point of view. Korzybski had written a protest letter to Hayakawa in 1946 (finally made public in 1990 in Supplementary V of *Alfred Korzybski: Collected Writings*). This was by no means the only bone of contention that Korzybski had with Hayakawa or other students of his work at the Society for General Semantics, which had started as a support group for Korzybski's Institute and which had turned into more of a rival, drawing the attention and funds of people interested in general semantics away from the Institute. Korzybski and his Institute colleagues and some of the Society leadership still hoped to find common ground for cooperation and perhaps even the eventual integration of the two organizations. But in the meantime, by 1948 Korzybski had decided to publicly stand his ground and to try to make as clear as possible important distinctions about his work that he felt had gotten distorted or forgotten. His adamancy about this seems quite apparent in the "Author's Note."

In 2010, general semantics, although not completely forgotten, certainly doesn't have the cachet it had in the late 1940s. Korzybski's problems with being superficially read or misread by some of his students may thus seem quaintly historical and mainly irrelevant now. But the superficial reading, misreading and non-reading of Korzybski's work certainly have persisted since his death in 1950. And ironically—as in Korzybski's day—korzybskian general semantics has often been most obscured by some individuals who call themselves 'general semanticists'. But there are new signs of interest. And there is no better way to clear aside the veil of distortions that have obscured Korzybski's work than to go directly to it. This new edition of *Selections from Science and Sanity*, originally produced by Guthrie Janssen with Korzybski's consultation and authorization, will make the prospect of reading Korzybski himself more inviting for those who might be put off by the "blue peril." Some readers will want to continue on to the peril itself. (Book Three, not excerpted in Selections, and which many readers avoid, contains many wonderfully-honed gems of insight about human evaluating.) In the final paragraph of the final page of Supplement III, "A Non-Aristotelian System and its Necessity For Rigour in Mathematics and Physics" (also not included in Selections but not to be missed), Korzybski summarized the scope and theme of his book:

> It is amusing to discover, in the twentieth century, that the quarrels between two lovers, two mathematicians, two nations, two economic systems . , [etc.] usually assumed insoluble in a 'finite period' should exhibit one mechanism—the semantic [evaluational] mechanism of identification—the discovery of which makes universal agreement possible, in mathematics and in life. (*Science and Sanity*, p. 761).

He wrote this in 1931. What if he was right? The implications for us in the twenty-first century are more than amusing.

Bruce I. Kodish
June 10, 2010
Pasadena, California

Fifth Edition of *Science and Sanity*, 1993

In the sixty years since Korzybski offered the first (not the last) non-Aristotelian system in *Science and Sanity* public reaction has been both enthusiastic and critical. What about Korzybski's work continues to generate such interest and activity? If it were so, as some critics have asserted, that `all' he did was to organize scattered insights, formulations and data into a system, that alone would have constituted a major achievement worthy of the gratitude of succeeding time-binders for centuries to come. Korzybski did that. He enunciated a system, incorporating aspects of but going beyond its predecessors, and proposed a methodology for making his system a living tool: *general semantics* (the name he selected), the first non-Aristotelian system *applied and made teachable*.

In addition, I can list here the following selected formulations and *points of view, emphases,* etc., which 1 consider to be original with Korzybski.[1]

1. *Time-binding; time-binding ethics*. Rejecting both theological and zoological definitions, Korzybski adopted a natural science, operational approach and defined humans by what they can be observed doing which differentiates them from other classes of life; he defined them as the *time binding class of life*, able to pass on knowledge from one generation to another over 'time.'[2]

Derived from this definition, which evaluates humans as a *naturally* cooperative class of life (the *mechanisms* of time-binding are descriptively social, cooperative), Korzybski postulated time-binding `ethics' - modes of behavior, *choices* appropriate to time-binding organisms.

2. Korzybski recognized that *language (symbolizing in general) constitutes the basic tool of time-binding*. Others before him had noted that *language*, in the complex human sense, was one of the distinguishing features of humans. What Korzybski *fully* recognized was the central, defining role of language. No language, no time-binding. If so, then *structures of languages* must be determinative for time-binding.

3. The *neurologically* focused formulation of the *process of abstracting*. *No one* before Korzybski had so thoroughly and unflinchingly specified the *process* by which humans build and evolve theories, do their mundane evaluating, thrill to `sunsets', etc. Korzybski's formulating of abstracting, particularly in the human realm, can constructively serve as a guide to on-going neuroscientific research.

4. As a function of the above, but deserving separate mention with the rigorously formulated notion of *orders* of abstracting, is the concurrent admonition that we should not confuse (identify) them. Given the hierarchical, sequential character of the nervous system (allowing, too, for horizontally related structures and parallel processing), it is inevitable that *results* along the way should manifest as (or `at') differing orders or levels of abstracting. These *results* are inevitable. That they would be *formulated* at a given historical moment is *not* inevitable. Korzybski did it in the 1920s, publishing his descriptions in their mature form in 1933.

5. *Consciousness of abstracting*. If human organisms-as-a-whole-cum-nervous systems/brains *abstract* as claimed above and described herein (pp. 369-451 and *passim*), surely *consciousness* of these events must be crucial for optimum human functioning. Animals (non-time-binders) abstract; but, so far as we know (1993), they do not *know* that they abstract. Indeed, many humans don't either - but they have the potentiality to do so. Korzybski recognized this, realized that *consciousness of abstracting* is essential for "fully functioning" humans, and made of it a primary goal of general semantics training.

6. The *structure of language*. Among Korzybski's most original formulations was the *multiordinal* character of many of the terms we most often use. He insisted that, with multiordinal terms, the `meaning' is strictly a function of the order or level of abstraction at which the term is used

and that its `meaning' is so context-driven that it doesn't `mean' *anything* definite until the context is specified or understood.

7. *Structure as the only `content' of knowledge.* This represents the height/depth of non-elementalism; that what used to be designated as `form' (structure) and `content' are so intimately related as to be, practically speaking, fused, that structure and `content' are functions of `each other'. Further, and more deeply, all we *can* ever know expresses a set or sets of *relationships* and, most fundamentally, a *relationship* ('singular') between the `known' and the knowing organism: the famous joint product of the observer-observed *structure*. "Structure is the only `content' of knowledge" may qualify as Korzybski's deepest expression of anti-essentialism. We can *not* know `essences', *things in themselves*; all we can know is what we know as *abstracting nervous systems*. Although we can on-goingly know *more*, we can not `transcend' ourselves as organisms that abstract.

8. *Semantic reactions; semantic reactions as evaluations.* Growing out of his awareness of the *transactive* character of human evaluating and wishing to correct for the elementalistic splitting involved in such terms as `meaning', `mental', `concept', `idea' and a legion of others, Korzybski consciously, deliberately formulated the term semantic reaction. It is central to his system.

9. *The mathematical notion of function as applied to the brain-language continuum.* Boldly grasping the neurophysiology of his day, Korzybski formulated what research increasingly finds: language is a function of (derives from, is invented by) brain; reciprocally, as a function of feedback mechanisms, brain is a function of (is modified by the electro-chemical structuring called) *language*.

10. *Neuro-semantic environments as environments.* The neurosemantic environment constitutes a fundamental environmental issue unique to humans.

11. Non-Aristotelian system *as system.* Korzybski had non-Aristotelian predecessors, as he well knew. What distinguishes his non-Aristotelian stance is *the degree of formulational consciousness* he brought to it, and the energetic courage with which he built it into a *system* -offered to his fellow humans as a better way to orient themselves.

12. *The Structural Differential as a model of the abstracting process* and a summary of general semantics. Korzybski realized the importance of *visualization* for human understanding. He knew, then, that to make some of the higher order, overarching *relationships* of his system accessible, *visible*, he must make a diagram, a model, a *map*, that people could *see* and *touch*. Thus the Structural Differential, a device for differentiating the structures of *abstracting*. As far as I know, this is the first structurally *appropriate* model of the abstracting process.

13. *Languages, formulational systems, etc., as maps and only maps of what they purport to represent.* This awareness led to the three premises (popularly expressed) of general semantics:

the map is *not* the territory

no map represents *all* of `its' presumed territory

maps are self-reflexive, i.e., we can map our maps indefinitely. Also, every map is *at least*, whatever else it may claim to map, a map of the map-maker: her/his assumptions, skills, world-view, etc.

By `maps' we should understand everything and anything that humans formulate - including this book and my present contributions, but also including (to take a few in alphabetical order), biology, Buddhism, Catholicism, chemistry, Evangelism, Freudianism, Hinduism, Islam, Judaism, Lutheranism, physics, Taoism, etc., etc...!

14. *Allness/non-allness as clear, to be dealt with, formulations*. If no map can represent *all* of `its' presumed territory, we need to eschew habitual use of the term `all' and its ancient philosophical correlates, *absolutes* of various kinds.

15. *Non-identity* and its derivatives, correlates, etc. At every turn in Korzybski's formulating we encounter his forthright challenge to the heart of Aristotelianism - *and* its non-Western, equally essentialist counterparts. "Whatever you say a thing is, it is *not*." This rejection of the `law of identity' ('everything is identical with itself) may be Korzybski's most controversial formulation. After all, Korzybski's treatment directly challenges the `Laws of Thought', revered for over two thousand years in the West and, differently expressed, in non-Western cultures. Korzybski's challenge is thus *planetary*. We `Westerners' can't (as some have tried) escape to the `East'. Identifications, confusions of orders of abstracting, are common to all human nervous systems we know of.

16. *Extension of Cassius Keyser's "Logical Fate"*: from premises, conclusions follow, *inexorably*. Korzybski recognized that conclusions constitute *behaviors, consequences, doings*, and that these are not merely logical derivatives but *psycho-logical* inevitabilities. If we want to change *behaviors*, we must first change the premises which gave birth to the behaviors. Korzybski's *strong version* of Keyser's restrictedly `logical' formulation was first adumbrated in Korzybski's paper, "Fate and Freedom" of 1923 and received its full expression in the "Foreword" (with M. Kendig) *to A Theory of Meaning Analyzed* in 1942, both available in the *Collected Writings*. Both expressions well antedate Thomas Kuhn's "paradigm shifts" and, more pointedly than Kuhn, formulate the behavioral implications of logical and philosophical systems.

17. *The circularity of knowledge* (spiral-character-in-'time'). Korzybski noted that our most `abstract' formulations are actually about nonverbal processes/events, and that how we formulate about these at a given date, how we talk to ourselves, through neural feedback mechanisms, relatively *determines* how we will subsequently abstract-formulate: healthfully if our abstracting is open, non-finalistic (non-absolute); pathologically if not.

18. *Electro-colloidal* (macro-molecular-biological) and related *processes*. Korzybski emphasized awareness of these as fundamental for understanding neuro-linguistic systems/organisms.

19. *Non-elementalism applied* to human organisms-as-a-whole-in-an environment. Some of Korzybski's predecessors in the study of language and human error may have pointed to what he labeled 'elementalism' (verbally splitting what cannot be split empirically) as a linguistically-embedded human habit, but none I know of had so thoroughly *built against it* and recommended replacing it with *habitual non-elementalism*. Korzybski's practical insistence that adopting non-elementalistic *procedures and terms* would benefit the humans (including scientists) who adopt them is original and, for him, urgent.

20. Extensions of *logics* (plural) as subsets of non-Aristotelian evaluating, including the limited usefulness (but *usefulness*) of Aristotelian logic.[3]

21. Epistemology as centered in *neuro-linguistic, neuro-semantic* issues. Korzybski built squarely on the neuroscience of his day and affirmed the fundamental importance of epistemology (the study of how we know what we *say* we know) as the *sine qua non* for any sound system upon which to organize our interactions with our children, students, friends, lovers, bosses, trees, animals, government - the `universe'. Becoming conscious of abstracting constitutes *applied* epistemology: *general semantics*.

22. *The recognition of and formulation of extensional and intensional orientations as orientations*. Here we see Korzybski at his most diagnostic and prognostic. Realizing that a person's epistemological-evaluational *style*, a person's *habitual way with evaluating determines* how life will go,

he recommends adoption of an extensional orientation, with its emphasis on `facts'. If a person is *over*-committed to verbal constructs, definitions, formulae, `conventional wisdom', etc., that person may be so trapped in those *a priori decisions* as to be unable to appropriately respond to new data from the non-verbal, not-yet-anticipated world. By definition, the *extensionally* oriented person, while remaining as articulate as any of her/his neighbors, is
habitually open to new data, is *habitually* able to say, "I don't know; let's see."

23. *Neuro-linguistic and neuro-semantic factors applied* to psychotherapeutic procedures and to the prevention of psycho-logical problems.

24. *Mathematics*. Korzybski's use of mathematical formulations and point of view qualifies as one of his most daring contributions.

25. *Science and mathematics as human behaviors*. Perhaps showing some Korzybskian influence (much of it has come to be `in the air'), writers on science and mathematics are increasingly addressing the human being who *does* science and/or mathematics. But Korzybski seems the first, to the *degree* that he did, to point to understanding these human behaviors as a necessary prerequisite or accompaniment to fully understanding science and mathematics as such. As Gaston Bachelard observed,

> The psychological and even physiological conditions of a non-Aristotelian logic have been resolutely faced in the great work of Count Alfred Korzybski, *Selections from Science and Sanity*.[4]

26. *Limitations of subject predicate languages* (modes of representation) when employed without consciousness of abstracting. Korzybski addresses this central formulation fully in his book.

27. Insistence on *relative* `invariance under transformation'. Korzybski was concerned that invariance of relations not be confused with `invariance' of processes.

28. *General uncertainty* (all statements merely probable in varying degrees) as an inevitable derivative of Korzybskian *abstracting, non-identity*, etc. Korzybski, drawing partly on his Polish milieu, anticipated and exceeded Heisenberg's mid-nineteen-twenties formulation of (restricted) uncertainty.

29. *The mechanism/machine-ism distinction*. This may seem too simple to list as an `original' or even major point. Yet it is vital, indicating as it does Korzybski's strong commitment to finding out how something *works* as opposed to vague, `spiritual' explanations.

Korzybski and some of his Institute successors who have worked to present *Korzybskian* general semantics have sometimes met this resistance: "I'm not a machine!" People trained in the myriad `intellectual', `mystical' in varying degrees systems/traditions they bring to seminars often react as if they fear to lose their `humanity' by being asked to consider the *mechanisms* which underpin or constitute their functioning. Korzybski took pains to explain that mechanism should not be confused with `machine-ism'. His concern for investigations at this level was bracing and central to his approach.

30. 'Infinite'-valued evaluating and semantic *methods* of science (*not* `content' of science or non-professional behavior of scientists at a given date) *as methods for sanity*. Thus the title of his book. General semanticists are obliged to evaluate, to analyze, criticize and sometimes reject the products of `science' at a given date. The approach is scientific, not scientistic.

31. *Predictability as the primary measure of the value of an Epistemological formulation*. Korzybski was by no means an 'anti-aesthete'. He was deeply sensitive to (and knowledgeable about) music, married a portrait painter, read literature (Conrad was a favorite) including poetry, and even liked to relax with a good detective story. But he insisted that, for life issues, beauty or cleverness or mere consistency (logical coherence, etc.) were not enough.

Korzybski offered his non-Aristotelian system with general semantics as its *modus operandi* as an on-going human acquisition, neg-entropic, ordering and self-correcting through and through, since it provides, self-reflexively, for its own reformulation, and assigns its users responsibility to do so should the need arise.

The above considerations have led me to the conclusion that Korzybski was not only a bold innovator, but also a brilliant synthesizer of available data into a coherent system. This system, when internalized and applied, can create a saner and more peaceful world, justifying the title of this book, *Science and Sanity*.

Robert P. Pula
September 1993

Notes

[1] For Korzybski as a system-builder, see Dr. Stuart A. Mayper, "The Place of Aristotelian Logic in Non-Aristotelian Evaluating: Einstein, Korzybski and Popper," *General Semantics Bulletin* No. 47, 1980, pp. 106-110. For discussions of the continuing appropriateness of `Korzybski's science', see Stuart A. Mayper, "Korzybski's Science and Today's Science," *General Semantics Bulletin*, No. 51, 1984, pp. 61-67; Barbara E. Wright, "The Heredity-Environment Continuum: Holistic Approaches at `One Point in Time' and `All Time'," *General Semantics Bulletin* No. 52, 1985, pp. 36-50; Russell Meyers, MD, "The Potentials of Neuro- Semantics for Modern Neuropsychology" (The 1985 Alfred Korzybski Memorial Lecture), *General Semantics Bulletin*, No. 54, 1989, pp. 13-59, and Jeffrey A. Mordkowitz, "Korzybski, Colloids and Molecular Biology: A View from 1985," *General Semantics Bulletin*, No. 55, 1990, pp. 86- 89. For a detailed updating in the neurosciences by a non-general semanticist which shows Korzybski's 1933 formulations as consistent with 1993 formulations, see "Part I" of Patricia Smith Churchland's *Neurophilosophy: Toward a Unified Science of the Mind/Brain*, MIT Press, 1986, pp. 14-235.)

[2] Korzybski, Alfred, *Manhood of Humanity*.

[3] See Dr. Mayper's paper: "The Place of Aristotelian Logic in Non-Aristotelian Evaluating. ..," listed above, and his earlier "Non-Aristotelian Foundations: Solid or Fluid?" *ETC.: A Review of General Semantics*, Vol. XVIII, No. 4, Feb., 1962, pp. 427-443.

[4] Bachelard, Gaston, *The Philosophy of No*. Translated from the French by G. C. Waterson. New York: Orion Press, 1968, p. 108.

Author's Note

These selections from *Science and Sanity: An Introduction to Non-Aristotelian Systems* and *General Semantics* were produced on the request of a number of teachers of General Semantics and study group leaders. They found that for some students the full text was too bulky or too expensive; yet they needed some fundamental textbook preserving the physico-mathematical approach. Originally I wrote *Selections from Science and Sanity* for scientists, teachers, and other leaders in our civilization. In my judgment all the material presented was necessary for them, but not as necessary for beginning students.

Personally I would be biased in making any "selections" from *Science and Sanity* and so had to rely on some teacher experienced with college and university students.

One such teacher, Guthrie E. Janssen, undertook the difficult task of making these selections. Following graduation from the University of Illinois in 1938 Mr. Janssen spent six years as an instructor of English and history in American schools in Egypt, particularly the American University at Cairo where he used *Science and Sanity* as a textbook with university students on the third year level. The following two years Mr. Janssen was war correspondent and broadcaster for the National Broadcasting Company, attached for a period to the United States Strategic Air Forces (B-29s). After travelling in some twenty-seven countries and broadcasting into the NBC network from Cairo, Athens, London, Manila, Tokyo, Shanghai, and from an airplane over Nagasaki, etc., and seeing results of the atomic bomb as one of the first ten Americans to enter Hiroshima, Mr. Janssen returned to this country and was granted a fellowship (donated by Robert K. Straus) for a year's study at the Institute of General Semantics. He produced these *Selections* as part of his working fellowship during 1946-1947.

I personally am most grateful to Guthrie Janssen for his considerable painstaking work, and to the Institute staff and others for their valuable suggestions and help in production. I wish to express my particular appreciation to M. Kendig, the Educational Director of the Institute; she urged for many years that such selections should be published, and gave valuable aid in bringing about its realization.

For teachers and students who will use this book I wish to include a forwarding concerning the fundamental confusion existing today about what the terms "semantics" and *General Semantics*[1] represent.

The original French *sémantique* was introduced into the literature by Bréal in 1897 in his *Essai de sémantique; science des significations*, which was translated into English in 1900 under the title *Semantics: Studies in the Science of Meaning*. Unfortunately the terms are not exactly equivalent in the different languages, and thus caused a confusion among the English-speaking people about the use of the term 'semantic' and 'semantics' which persists up to today. Sémantique deals with a branch of philology and the historical change of significance ('meaning'). Lady Welby somehow felt that difference in implication and formulated a more organismal theory under the name of 'Significs'. The Significs International Movement in the Netherlands is still carrying out this work, under the leadership of mathematicians such as Brouwer (the founder of the Intuitional School in Mathematics) and logicians, epistemologists, psychologicians, etc.

Both disciplines labelled by those terms were not non-elementalistic enough, and so different researchers attempted further elaborations and amplifications under various old or new terms such as 'semasiology', 'semiosis', 'semiotic', etc.

As to the relationship between those disciplines, Lady Welby wrote in the eleventh edition of the *Encyclopedia Britannica*, 'Semantics may . . . be described as the application of Significs within strictly philological limits.'

In his *Introduction to Semantics* (p. 9) Rudolf Carnap says, 'If in an investigation explicit refer-ence is made to the . . . user of a language [from a businesslike, practical point of view] then we assign it to the field of pragmatics [from the Greek *pragmatikos*, deed, business, act, etc.] . . . If we abstract from the user of the language [i.e., disregard the person] and analyze only the expressions and their designata [referents?] we are in the field of semantics. And if, finally, we abstract from the designata also and analyze only the relations between the expressions, we are in (logical) syntax. The whole science of language, consisting of the three parts mentioned, is called semiotic.[2]'

Obviously such a 'whole science of language' consisting of 'pragmatics', 'semantics', and 'logic', which is called 'semiotic', disregards the inner reactions of the individual person, and so elimi-nates the *possibility of evaluation* as a living issue with a living and individual, which is the main aim of *General Semantics*.

Charles Morris says explicitly that 'Semiotic is not then a "theory of value".' Of 'Semantics' he writes, 'That branch of semiotic which studies the signification of signs.' (Signs, *Language and Behavior*, pp. 80 and 353). Of my work he says, 'The work of A. Korzybski and his followers is psy-cho-biological in orientation . . . aiming to protect the individual against exploitation by others and by himself,' (p. 283), in other words, dealing with the inner life of the individual, on the silent (non-verbal) levels.

From what was said here it is obvious that my work in General Semantics has nothing to do with the above-mentioned disciplines, although I know and respect the works of the corresponding investigators in those fields, with their stated limitations.

Even in the index of *Science and Sanity* the word 'semantics' does not appear except as 'Se-mantics, General'. I use 'semantic' there only as an adjective with other words, in the sense of 'evalu-ational', such as 'semantic aphasia', 'semantic blockage', 'semantic reactions', etc. I selected the term 'General Semantics' for an empirical natural science of non-elementalistic *evaluation*, a theory of values.

If I had not known of the work done in *Sémantique*, Significs, etc., I would have labeled my work by another name, but my system would have remained fundamentally unaltered. Thus, my papers before the International Mathematical Congress in Toronto in 1924, before the Washington Society for Nervous and Mental Diseases in 1925, and before the Washington Psychopathic Society in 1926 outlined practically my whole system before I became familiar with the works of Bréal, Lady Welby, et al. The word 'semantic' does not appear in those papers at all, and my work is called 'Time-binding, the General Theory', which remains as important as ever today. I also coined, I believe originally, the term 'human engineering', but since the publication of my *Manhood of Human-ity: The Science and Art of Human Engineering* in 1921 that term has been so abused that I had to abandon it, and actually had to hunt for another term. 'Semantics', 'significs', etc., were unusable, as they did not even touch my field. From a time-binding point of view, and in fairness to the efforts of others, I coined the term 'General Semantics' on the assumption that intelligent laymen will be able to discriminate between 'semantics' and 'General Semantics', as mathematicians are able to discrimi-nate between the Cartesian system and the vector, tensor, etc., calculuses as different disciplines, in the process of mathematical evolution.[3]

I selected it also for historical continuity, as the problems on the non-verbal levels outside or inside our skins are present with us and real, no matter whether their relations to the verbal levels were solved by my predecessors and contemporaries or not. The term 'General Semantics' seemed most appropriate to me because of the derivation from the Greek *semainein*, 'to mean', 'to signify'. A theory of evaluation seemed to follow naturally in an evolutionary sense from 1) 'meaning"' to 2) 'signification' to 3) *evaluation, if we take into account the individual*, not divorcing him from his reactions, nor from his *neuro*-linguistic and neuro-semantic environments. Thus we allocate him in a *plenum* of some values, no matter what, and a *plenum* of language, which may be used to inform, or misinform by commission and/or omission, deceiving the individual himself and/or others. With such problems, without exception, the individual has to cope to be human at all. That's what I learned from the theory of time-binding and what I tried to convey to others through General Semantics and psycho-biological non-Aristotelian considerations.

I showed several years ago that theories of 'meaning' are humanly impossible, as they do not take into consideration *undefined* terms, which label only the silent levels of non-verbal experiences, etc.[4] Confusion between non-verbal silent levels, and verbal levels, due to lack of consciousness of abstracting, leads inevitably to insidious identifications (mis-evaluations) of these different levels. Primitivism, infantilism, formalism, academic stupidities, un-sanity, and other types of pathological reactions, must then follow.

The words 'semantic' and 'semantics' are today commonly used even in newspapers and magazines mostly in the sense of 'meaning'. Important scientists, mathematicians and physicists included, also use these words, mostly in that sense. Many of them know something about General Semantics, and if they mention my work at all, they say explicitly that they use the term 'semantic' in an entirely different sense that I use the term 'General Semantics', and they are exactly correct.

The more my researches advanced, the more it became obvious that deeper studies in many branches of science were necessary. I had to investigate further hidden silent assumptions. Finally it became clear that nothing short of a *methodological syntheses* of mathematics and modern empirical sciences would suffice for a grand theory of *values*. This synthesis turned out to be (although it was not planned as such) a non-Aristotelian system, the first so far to be formulated. Today it becomes impossible to separate General Semantics and this Non-Aristotelian System. One follows from the other, and vice versa, General Semantics, being the *modus operandi* and foundation of the system.

As the center for training in these non-Aristotelian methods, the Institute of General Semantics was incorporated in Chicago in 1938. In the summer of 1946 the Institute moved to Lakeville, Connecticut where its original program is being carried on. The rapid spread of interest in our work, by now on all continents, has indicated the need for the new methods set forth here. I must stress that General Semantics gives no panaceas, but experience shows that when the methods of General Semantics are *applied*, the results are usually beneficial, whether in law, medicine, business, etc., education on all levels, or personal interrelationships, be they family, community, national, or international. If the methods are not applied, but merely talked about, no results can be expected. Perhaps the most telling applications were those on the battlefields of World War II, as reported by members of the armed forces, including psychiatrists on all fronts, and especially by Dr. Douglas M. Kelley, formerly Lieutenant Colonel in the Medical Corps[5], who reports in part as follows:

> General semantics, as a modern scientific method, offers techniques which are of extreme value both in prevention and cure of such [pathological] reactive patterns. In my experience with over seven thousand cases in the European Theater of Operations, these basic principles were daily employed as methods of group psychotherapy and as methods of psychiatric prevention. It is obvious that the earlier the case is treated the better the prognosis, and consequently hundreds of battalion-aid surgeons were trained in principles of general semantics. These principles were applied (as individual therapies and as group therapies) at every treatment level from the forward area to the rear-most echelon, in front-line aid stations, in exhaustion centers and in general hospitals. That they were employed with success is demonstrated by the fact that psychiatric evacuations from the European Theater were

held to a minimum.

It is not generally realized that with human progress, the complexities and difficulties in the world increase following an exponential function of 'time', with indefinitely accelerating accelerations. I am deeply convinced that these problems cannot be solved at all unless we boldly search for and revise our antiquated notions about the 'nature of man' and apply modern extensional methods toward their solution. Let us also remember that the methods of exact sciences disregard national boundaries, and so the extensional methods and devices of General Semantics can be applied to all existing languages, with deep psycho-logical effects on the users and through them on their countrymen. Thus the world can gain an international common denominator for inter-communication, mutual understanding, and eventual agreement.

A.K.

Lakeville, Connecticut

February, 1948

Additional Note: As this was going to press a new paper by Allen Walker Read of New York, to be published soon, came to my attention. One paragraph in particular represents such an excellent, terse, historical statement of how I came to introduce the term 'General Semantics', that I asked for, and received, Mr. Read's kind permission to reproduce it here:

The great popular vogue of the word *semantics* can be traced to the ferment caused by the works of Alfred Korzybski. In 1928, in the first draft of his *Science and Sanity*, he did not make use of *semantics*, *general semantics*, or *semantic reaction* at all. But . . . he was keeping in touch with the developments among Polish mathematicians, and he was particularly impressed with their work upon attending the "Congrés des mathématiciens des pays Slaves" in Warsaw in 1929. In 1931, in a paper given before the American Mathematical Society at New Orleans, Louisiana, he presented material on "the *restricted semantic* school represented by Chwistek and his pupils, which is characterized mostly by the semantic approach." ("A Non-Aristotelian System and its Necessity for Rigour in Mathematics and Physics," printed in *Science and Sanity*, pp. 747-761; quotation, p. 748). He announced that he was using the term "general semantics" for his own study (Ibid., p. 749). Before this he has called his work "Time-binding, the general theory."), and that his researches had resulted "in the discovery of a general semantic mechanism underlying human behavior, many new interrelations and formulations, culminating in a [Non-Aristotelian]-system." (Ibid., p. 750). Thus the background of Korzybski's usage is found in the Polish logicians, though some of his followers have erroneously associated it with the antiquarianism of Bréal, Ernest Weekley, and popular writers on the "the glamour of world study."

A.K

[1] For example in the *Dictionary of World Literature* 'General Semantics' appears under the term Semantics. The two disciplines are confused, and even my 'extensional devices' are called 'semantic devices', whereas such a thing does not exist.

In Vol. III, No. 4 of *ETC.*, there appears a five-page glossary of terms used in General Semantics, all of them fully explained in *Science and Sanity*. Practically every 'definition' misses the main point and trend of my work. For instance, what is said in the glossary about the use of the term "semantic" in my 'General Semantics', i.e., in a new theory of values, is entirely misleading. Such initial errors lead automatically to further more aggravated misinterpretations. It would not be an overstatement to say that definitions of 'semantics' and '*General Semantics*' and other terms printed in this glossary must be considered as seriously misinforming the public. The most recent example is an article on 'Semantics, General Semantics' in *Ten Eventful Years*, an *Encyclopedia Britannica* publication, which considerably increased the confusion. It is not even mentioned that 'semantics' is a branch of philology, nor is there any clarifying discrimination made between the noun 'semantics' and the adjective 'semantic'. Moreover, it has many misstatements and even falsifications of my work and the work of others, and some statements make no sense.)

[2] Carnap is an important constructive worker in the field of 'logic', representative of the Viennese school. However, even he adds to the confusion between 'semantics' and 'General Semantics'. Carnap was eventually entitled to use the term 'semantics', as he may be considered to deal with the philology of logical and mathematical languages. He seems acquainted with my *Science and Sanity: An Introduction to Non-Aristotelian Systems and General Semantics*, which was published in 1933. Yet in his book published in 1942 he uses the term 'General Semantics' when he actually wants to say 'generalized semantics', and so adds to the linguistic chaos. Further dangers are still ahead, as Carnap announces on page ix a series of books, 'Studies in Semantics', where this misuse of language and terms may be propagated. The term 'General Semantics' was introduced in 1933 as a technical term in an empirical natural science of evaluation which deals with living human reactions and has nothing to do with 'logic' or mathematics as such.

[3] Historically it is interesting to note that the original manuscript of *Science and Sanity* did not contain the word 'semantics' or 'semantic'.

[4] See Alfred Korzybski and M. Kendig, 'Foreword' to A Theory of Meaning Analyzed, *General Semantics Monographs*, No. III, Institute of General Semantics, 1942.

[5] Chief Consultant in Clinical Psychology and Assistant Consultant in Psychiatry to the European Theater of Operations; also Chief Psychiatrist in charge of the prisoners at Nuremberg. Author of 22 *Cells in Nuremberg*, Greenberg, New York, 1947.

[6] Korzybski, A. A Veteran's Re-Adjustment and Extensional Methods. *ETC: A Review of General Semantics*, Vol. III, No. 4. See also

Saunders, Captain James, USN (Ret.). Memorandum. *Training of Officers for the Naval Service: Hearings Before the Committee on Naval Affairs*, United States Senate, June 12 and 13, 1946, pp. 55-57. U.S. Government Printing Office, Washington, D.C., 1946.

Selections from Science and Sanity

Note to Teachers and Group Leaders

The criterion for making these *Selections* was the question, 'On the basis of experience with students of General Semantics, what portions of *Science and Sanity* seem most appropriate and important for classes on the beginning level?' They represent nothing more pretentious than this.

This volume is printed as a limited, experimental edition for use in the classroom or by study groups preliminary to the full text.

It is assumed that the teacher or group leader is an advanced student, thoroughly familiar with the full text, including quotations and bibliography.

Except for topics 1 and 19 the sequence follows *Science and Sanity*. Topic 1 contains selections from Chapter I, Chapter III, Preface to the First Edition, and Introduction to the Second Edition. . . .

Students and teachers are reminded that Book I (in general, topics 1 through 11) represents 'A General Survey of Non-Aristotelian Factors', an introduction to the new synthesis presented in Book II (topics 12 through 19). Since Book I is usually more difficult for the beginner than Book II the student should observe the old maxim that prefaces Introduction to the Second Edition: 'When in perplexity, read on.'

Because material in them was treated at least briefly in selections from other chapters, the following were omitted entirely:

VII	Linguistic Revision
VIII	General Epistemological
IX	Colloidal Behaviour
X	The 'Organism-as-a-Whole'
XIV	On the Notion of Infinity
XV	The "Infinitesimal" and "Cause and Effect"
XVI	On the Existence of Relations
XXII	On "Inhibition"
XXII	On Conditional Reactions of Higher Orders and Psychiatry

Book III, which contains 'Additional Structural Data about Languages and the Empirical World' has been omitted entirely since its full text is not generally considered absolutely essential for the beginning student. The teacher or group leader should have studied it in detail. It is assumed he will use Book III to supply additional non-Aristotelian data when needed and that he is also familiar in general with developments in the non-Aristotelian direction in physico-mathematical and other fields since 1933.

Most of the 265 quotations have been omitted. The few included indicate, like others in the full text, that much old and current genuine wisdom in the world has not yet been applied. Korzybski points out it *could not be applied* until a new synthesis was made, with methods and techniques for training human nervous systems. The new synthesis, first presented in *Science and Sanity*, turned out to be a non-Aristotelian system, with General Semantics as *modus operandi*. (For sources of quotations refer to Bibliography in the full text; number is given in parentheses at the end of each quotation.)

Selections from Introduction to the Second Edition (1941) and Preface to the First Edition have been inserted among those from the main text, which has not changed since publication of the First Edition in 1933. . . . Numbered footnotes have been omitted. The student is reminded that incidental mentions of page numbers, numbers of Parts, etc. refer to the full text.

In connection with 'conditional reflexes' (p. 157) teachers are refereed to full text p. 329 in which Korzybski explains greater adequacy of the term 'reaction' instead of 'reflex'.

I wish to thank the following teachers in universities and professional schools who have used *Science and Sanity* with their classes in various subjects for valuable suggestions: O.R. Bontrager, State Teachers College, California, Penna.; David. D. Eitzen, University of Southern California, Los Angeles; Gordon K. Haist, Rose Polytechnic Institute, Terre Haute, Indiana; Wendell Johnson, State University of Iowa, Iowa City; Douglas M. Kelley, M.D., Bowman Gray School of Medicine, Winston-Salem, N.C.; Irving J. Lee, Northwestern University, Evanston, Illinois; Elwood Murray, University of Denver; and Bess Sondel, University of Chicago. I wish also to thank R.H. Baugh and Ralph Hamilton of the Institute staff and Robin Skynner of London, England, for their help in preparation of copy.

Teachers and Group Leaders who will use this edition are invited offer suggestions for future editions, especially on parts they feel might be omitted, other parts that should be included, on general sequence, organization, etc.

G.E.J.

QUOTATIONS

Being myself a remarkably stupid fellow, I have had to unteach myself the difficulties, and now beg to present to my fellow fools the parts that are not hard. Master these thoroughly, and the rest will follow. What one fool can do, another can. *(510)*

SILVANUS P. THOMPSON

Of all men, Aristotle is the one of whom his followers have worshipped his defects as well as his excellencies: which is what he himself never did to any man living or dead; indeed, he has been accused of the contrary fault. *(354)*

AUGUSTUS DE MORGAN

The every-day language reeks with philosophies . . . It shatters at every touch of advancing knowledge. At its heart lies paradox.

The language of mathematics, on the contrary, stands and grows in firmness. It gives service to men beyond all other language. *(25)*

ARTHUR F. BENTLEY

These difficulties suggest to my mind some such possibility as this: that every language has, as Mr. Wittgenstein says a structure concerning which, in the language, nothing can be said, but that there may be another language dealing with the structure of the first language, and having itself a new structure, and that to this hierarchy of languages there may be no limit. Mr. Wittgenstein would of course reply that his whole theory is applicable unchanged to the totality of such languages. The only retort would be to deny that there is any such totality. *(456)*

BERTRAND RUSSELL

The wretched monosyllable "all" has caused mathematicians more trouble than all the rest of the dictionary. *(23)*

E.T. BELL

If he contend, as sometimes he will contend, that he has defined all his terms and proved all his propositions, then either he is a performer of logical miracles or he is an ass; and, as you know, logical miracles are impossible. *(264)*

CASSIUS J. KEYSER

The firm determination to submit to experiment is not enough; there are still dangerous hypotheses; first, and above all, those which are tacit and unconscious. Since we take them without knowing it, we are powerless to abandon them. *(417)*

H. POINCARÉ

The current accounts of perception are the stronghold of modern metaphysical difficulties. They have their origin in the same misunderstanding which led to the incubus of the substance-quality categories. The Greeks looked at a stone, and perceived that it was grey. They Greeks were ignorant of modern physics; but modern philosophers discuss perception in terms of categories derived from the Greeks. *(574)*

A.N. WHITEHEAD

To sum up, we can say that the fundamental grammatical categories, universal to all human languages, can be understood only with reference to the pragmatic Weltanschauung of primitive man, and that, through the use of Language, the barbarous primitive categories must have deeply influenced the later philosophies of mankind. *(332)*

B. MALINOWSKI

Consciousness is the feeling of negation: in the perception of 'the stone as grey,' such feeling is in barest germ; in the perception of 'the stone as not grey,' such feeling is full development. Thus the negative perception is the triumph of consciousness. *(578)*

A. N. WHITEHEAD

The little word is has its tragedies; it marries and identifies different things with the greatest innocence; and yet no two are ever identical, and if therein lies the charm of wedding them and calling them one, therein too lies the danger. Whenever I use the word is, except in sheer tautology, I deeply misuse it; and when I discover my error, the world seems to fall asunder and the members of my family no longer know one another. *(461)*

G. SANTAYANA

The supposition of common sense and naïve realism, that we see the actual physical object, is very hard to reconcile with the scientific view that our perception occurs somewhat later than the emission of light by the object; and this difficulty is not overcome by the fact that the time involved, like the notorious baby, is a very little one. *(457)*

BERTRAND RUSSELL

Neither the authority of man alone nor the authority of fact alone is sufficient. The universe, as known to us, is a joint phenomenon of the observer and the observed; and every process of discovery in natural science or in other branches of human knowledge will acquire its best excellence when it is in accordance with this fundamental principle. *(82)*

R.D. CARMICHAEL

The empiricist . . . thinks he believes only what he sees, but he is much better at believing than at seeing. *(461)*

G. SANTAYANA

For, beyond the bounds of science, too, objective and relative reflection is a gain, a release from prejudice, a liberation of the spirit from standards whose claim to absolute validity melts away before the critical judgment of the relativist. *(45)*

MAX BORN

The layman, the 'practical' man, the man in the street says, What is that to me? The answer is positive and weighty. Our life is entirely dependent on the established doctrines of ethics, sociology, political economy, government, law, medical science, etc. This affects everyone consciously or unconsciously, the man in the street in the first place, because he is the most defenseless. *(280)*

A.K.

The only justification for our concepts and system of concepts is that they serve to represent the complex of our experiences; beyond this they have no legitimacy. I am convinced that the philosophers have had a harmful effect upon the progress of scientific thinking in removing certain fundamental concepts from the domain of empiricism, where they are under our control, to the intangible heights of the a priori. *(152)*

A. EINSTEIN

Common, do what it will, cannot avoid being surprised occasionally. The object of science is to spare it this emotion and create mental habits which shall be in such close accord with the habits of the world as to secure that nothing shall be unexpected. *(457)*

BERTRAND RUSSELL

In sum, *all* the scientist creates in a fact is the *language* in which he enunciates it. *(417)*

H. POINCARÉ

A civilization which cannot burst through its current abstractions is doomed to sterility after a very limited period of progress. *(575)*

A.N. WHITEHEAD

. . . almost any idea which jogs you out of your current abstractions may be better than nothing. *(575)*

A.N. WHITEHEAD

But it does not seem a profitable procedure to make odd noises on the off-chance that posterity will find a significance to attribute to them. *(149)*

A.S. EDDINGTON

The Dormouse...went on: "—that begins with an M, such as mousetraps, and the moon, and memory, and muchess— you know you say things are 'much of a muchness'—did you ever see such a thing as a drawing of a muchness!"
"Really, now you ask me," said Alice, very much confused, "I don't think—"
"Then you shouldn't talk," said the Hatter.** *4.1212*

LEWIS CARROLL

What can be shown cannot be said. *(590)*

L. WITTGENSTEIN

**Alice in Wonderland

Selections from Science and Sanity

Abbreviations

[Editor's Note: It was not feasible to reproduce the specialized null-A, null-E and null-N characters that Korzybski used in this edition, so they have been replaced with [non-A], [non-E], and [non-N] respectively.]

To make issues sharper, some words will be repeated so often that I abbreviate them as follows:

Abbreviation	Stands for	Abbreviation	Stands for
A	Aristotelian	[non-N]	non-Newtonian
[non-A]	non-Aristotelian	el	elementalistic
E	Euclidean	non-el	non-elementalistic
[non-E],	Non-Euclidean	m.o or (m.o)	multiordinal
N	Newtonian	s.r. or (s.r)	Semantic reactions, both singular and plural

In some instances, for special emphasis, the words will be spelled in full.

A system, being extensional, requires the enumeration of long lists of names, which, in principle, cannot be exhausted. Under such conditions, I have to list a few representatives followed by an "etc." or its equivalents. As the extensional method is characteristic of a   treatment, the expression "etc." occurs so often as to necessitate a special extensional punctuation whenever the period does not indicate another abbreviation, as follows:

Abbreviation	Stands For	Abbreviation	Stands for
.,	etc.,	.:	etc.:
,. .	,etc.	.?	etc.?
.;	etc.;	.!	etc.!

1. INTRODUCTORY

A Science of Man

The present enquiry originated in my attempt to build a science of man. The first task was to define man scientifically in non-elementalistic, functional terms. I accomplished that in my book *Manhood of Humanity* (New York, E. P. Dutton & Co.), and in it I called the special characteristic which sharply distinguishes man from animal the time-binding characteristic.

In the present volume I undertake the investigation of the mechanism of time-binding. The results are quite unexpected. We discover that there is a sharp difference between the nervous reactions of animal and man, and that judging by this criterion, nearly all of us, even now, copy animals in our nervous responses, which copying leads to the general state of un-sanity reflected in our private and public lives, institutions and systems. By this discovery the whole situation is radically changed. If we copy animals in our nervous responses through the lack of knowledge of what the appropriate responses of the human nervous system should be, we can stop doing so, both individually and collectively, and we are thus led to the formulation of a first positive theory of sanity.

The old dictum that we 'are' animals leaves us hopeless, but if we merely *copy* animals in our nervous responses, we can stop it, and the hopeless becomes very hopeful, provided we can discover a *physiological* difference in these reactions. Thus we are provided with a definite and promising program for an investigation.

In my *Manhood of Humanity*, it is shown that the canons of what we call 'civilization' or 'civilizations' are based on animalistic generalizations taken from the obvious facts of the lives of cows, horses, dogs, pigs., and applied to man. Of course, such generalizations started with *insufficient data*. The generalizations had to be primitive, superficial; and when applied in practice, periodical collapses were certain to follow. No bridge would stand or could even be built, if we tried to apply rules of surfaces to volumes. The rules or generalizations in the two cases are different, and so the results of such primitive semantic confusion must be disastrous to all of us.

The present enquiry began with the investigation of the characteristic difference between animal and man; namely, the mechanism of time-binding. This analysis, because of the different structure of the language used, had to be carried on independently and anew. The results are, in many instances, new, unexpected even by myself, and they show unmistakably that, to a large extent, even now we nearly all copy animals in our nervous processes. Investigation further shows that such nervous reactions in man lead to non-survival, pathological states of general *infantilism*, infantile private and public behaviour, infantile institutions, infantile 'civilizations' founded on strife, fights, brute competitions., these being supposedly the 'natural' expression of 'human nature', as different commercialists and their assistants, the militarists and priests, would have us believe.

As always in human affairs, in contrast to those of animals, the issues are circular. Our rulers, who rule our symbols, and so rule a symbolic class of life, impose their own infantilism on our institutions, educational methods, and doctrines. This leads to nervous maladjustment of the incoming generations which, being born into, are forced to develop under the un-natural (for man) semantic conditions imposed on them. In turn, they produce leaders afflicted with the old animalistic limitations. The vicious circle is completed; it results in a general state of human un-sanity, reflected again in our institutions. And so it goes, on and on.

At first, such a discovery is shocking. On second consideration, however, it seems natural that the human race, being relatively so recent, and having passed through different low levels of development, should misunderstand structurally their human status, should misuse their nervous system,. The present work, which began as the 'Manhood of Humanity' turned out to be the 'Adulthood of Humanity', for it discloses a *psychophysiological* mechanism of infantilism, and so points toward its prevention and to adulthood.

The term 'infantilism' is often used in psychiatry. No one who has had any experience with the 'mentally' ill, and studied them, can miss the fact that they always exhibit some infantile symptoms. Semantic characteristics, cannot be a fully adjusted individual, and usually wrecks his own and other persons' lives.

In the present investigation, we have discovered and formulated a definite psychophysiological mechanism which is to be found in all cases of 'mental' ills, infantilism, and the so-called 'normal' man. The differences between such neural disturbances in different individuals turn out to vary only in degree, and as they resemble closely the nervous responses of animals, which are necessarily regressive for man, we must conclude that, generally, we do not use our nervous system properly, and that we have not, as yet, entirely emerged from a very primitive semantic stage of development, in spite of our technical achievements.

Indeed, experience shows that the more technically developed a nation or race is, the more cruel, ruthless, predatory, and commercialized its systems tend to become. These tendencies, in turn, colour and vitiate international, national, capital-labour, and even family, relations.

Is, then, the application of science at fault? No, the real difficulty lies in the fact that different primitive, animalistic, un-revised doctrines and creeds with corresponding s.r have *not* advanced in an equal ratio with the technical achievements. When we analyse these creeds semantically, we find them to be based on structural assumptions which are false to facts, but which are strictly connected with the unrevised structure of the primitive language, all of which is the more dangerous because it works unconsciously.

When we study comparatively the nervous responses of animals and man, the above issues become quite clear, and we discover a definite psychophysiological mechanism which marks this difference. That the above has not been already formulated in a workable way is obviously due to the fact that the *structure* of the old language successfully prevented the discovery of these differences, and, indeed, has been largely responsible for these human semantic disturbances. Similarly, in the present [non-A]-system, the language of a new and modern structure, as exemplified by terms such as 'time-binding', 'orders of abstractions', 'multiordinal terms', 'semantic reactions'., led automatically to the disclosure of the mechanism, pointing the way toward the means of control of a special therapeutic and preventive character.

The net results are, in the meantime, very promising. Investigation shows that, in general, the issues raised are mostly *linguistic*, and that, in particular, they are based on the analysis of the *structure* of languages in connection with s.r. All statements, therefore, which are made in this work are about empirical facts, language and its structure. We deal with an obvious and well-known inherent psychophysiological function of the human organism, and, therefore, all statements can be readily verified or eventually corrected and refined, allowing easy application, and automatically eliminating primitive mythologies and s.r.

In 1933, scientific opinion is divided as to whether we need more science or less science. Some prominent men even suggest that scientists should take a vacation and let the rest of mankind catch up with their achievements. There seems no doubt that the discrepancy between human adjustments and the advances of science is becoming alarming. Is, then, such a suggestion justified?

The answer depends on the *assumptions* underlying such opinions. If humans, as such, have reached the limit of their nervous development, and if the scientific study of man, as man, should positively disclose these limitations, then such a conclusion would be justified. But is this the case?

The present investigation shows most emphatically that this is not the case. All sciences have progressed exclusively because they have succeeded in establishing their own [non-A] languages. For instance, a science of thermodynamics could not have been built on the terms of 'cold' and 'warm'. Another language, one of relations and structure, was needed; and, once this was produced, a science was born and progress secured. Could *modern* mathematics be built on the Roman notation for numbers —I, II, III, IV, V.? No, it could not. The simplest and most child-like arithmetic was so difficult as to require an expert; and all progress was very effectively hampered by the symbolism adopted. History shows that only since the unknown Hindu discovered the most revolutionary and modern principle of *positional notation*—1, 10, 100, 1000., modern mathematics has become possible. Every child today is more skillful in his arithmetics than the experts of those days. Incidentally, let us notice that positional notation has a definite *structure*.

Have we ever attempted anything similar in the study of man? As-a-whole? In all his activities? Again, most emphatically, No! We have never studied man-as-a-whole scientifically. If we make an attempt, such as the present one, for instance, we discover the astonishing, yet simple, fact that, even now, we all copy animals in our nervous responses, although these can be brought to the human level if the difference in the mechanism of responses is discovered and formulated.

Once this is understood, we must face another necessity. To abolish the discrepancy between the advancement of science and the power of adjustment of man, we must first establish the science of man-as-a-whole, embracing *all* his activities, science, mathematics and 'mental' ills *not* excluded. Such an analysis would help us to discover the above-mentioned difference in responses, and the s.r in man would acquire new significance.

If the present work has accomplished nothing more than to suggest such possibilities, I am satisfied. Others, I hope, will succeed where I may have failed. Under such conditions, the only feasible resort is to produce a science of man, and thus have not less, but more, science, ultimately covering all fields of human endeavour, and thereby putting a stop to the animalistic nervous reactions, so vicious in their effects on man.

General on Identification

The general character of the present work is perhaps best indicated by the two following analogies. It is well known that for the working of any machine some lubricant is needed. Without expressing any judgement about the present 'machine age', we have to admit that technically it is very advanced, and that without this advancement many scientific investigations necessitating very refined instruments would be impossible. Let us assume that mankind never had at its disposal a clean lubricant, but that existing lubricants always contained emery sand, the presence of which escaped our notice. Under such conditions, existing technical developments, with all their consequences, would be impossible. Any machine would last only a few weeks or months instead of many years, making the prices of machines and the cost of their utilization entirely prohibitive. Technical development would thus be retarded for many centuries. Let us now assume that somebody were to discover a simple means for the elimination of emery from the lubricants; at once the present technical developments would become possible, and be gradually accomplished.

Something similar has occurred in our human affairs. Technically we are very advanced, but the elementalistic premises underlying our human relations, practically since Aristotle, have not changed at all. The present investigation reveals that in the functioning of our nervous systems a special harmful factor is involved, a 'lubricant with emery' so to speak, which retards the development of sane human relations and prevents general sanity. It turns out that in the structure of our languages, methods, 'habits of thought', orientations, etc., we preserve delusional, psychopathological factors. These are in no way inevitable, as will be shown, but can be easily eliminated by special training, therapeutic in effect, and consequently of educational preventive value. This 'emery' in the nervous system I call identification. It involves deeply rooted 'principles' which are invariably false to facts and so our orientations based on them cannot lead to adjustment and sanity.

A medical analogy here suggests itself. We find a peculiar parallel between identification and infectious diseases. History proves that under primitive conditions infectious diseases cannot be controlled. They spread rapidly, sometimes killing off more than half of the affected population. The infectious agent may be transmitted either directly, or through rats, insects, etc. With the advance of science, we are able to control the disease, and various important preventive methods, such as sanitation, vaccination, etc., are at our disposal.

Identification appears also as something 'infectious', for it is transmitted directly or indirectly from parents and teachers to the child by the mechanism and structure of language, by established and inherited 'habits of thought', by rules for life-orientation, etc. There are also large numbers of men and women who make a profession of spreading the disease. Identification makes general sanity and complete adjustment impossible. Training in non-identity plays a therapeutic role with adults. The degree of recovery depends on many factors, such as the age of the individual, the severity of the 'infection', the diligence in training in non-identity, etc. With children the training in non-identity is extremely simple. It plays the role both of sanitation and of the equally simple and effective preventive vaccination.

As in infectious diseases, certain individuals, although living in infected territory, are somehow immune to this, disease. Others are hopelessly susceptible.

The present work is written on the level of the average intelligent layman, because before we can train -children in non-identity by preventive education, parents and teachers must have a handbook for their own guidance. It is not claimed that a milleniurn is at hand, far from it; yet it seems imperative that the *neuro-psycho-logical* factors which make general sanity impossible should be eliminated.

I have prefaced the parts of the work and the chapters with a large number of important quotations. I have done so to make the reader aware that, on the one hand, there is already afloat in the 'universe of discourse' a great deal of genuine knowledge and wisdom, and that, on the other hand, this wisdom is not generally applied and, to a large, extent, cannot be applied as long as we fail to build a simple system based on the complete elimination of the pathological factors.

A system, in the present sense, represents a complex whole of coordinated doctrines resulting in methodological rules and principles of procedure which affect the orientation by which we act and live. Any system involves an enormous number of assumptions, presuppositions, etc., which, in the main, are not obvious but operate unconsciously. As such, they are extremely dangerous, because should it happen that some of these unconscious presuppositions are false to facts, our whole life orientation would be vitiated by these unconscious delusional factors, with the necessary result of harmful behaviour and maladjustment. No system has ever been fully investigated as to its underlying unconscious presuppositions. Every system is expressed in some language of some structure, which is based in turn on silent presuppositions, and ultimately reflects and reinforces those presuppositions on and in the system. This connection is very close and allows us to investigate a system to a large extent by a linguistic structural analysis.

The system by which the white race lives, suffers, 'prospers', starves, and dies today is not in a strict sense an aristotelian system. Aristotle had far too much of the sense of actualities for that. It represents, however, a system formulated by those who, for nearly two thousand years since Aristotle, have controlled our knowledge and methods of orientations, and who, for purposes of their own, selected what today appears as the worst from Aristotle and the worst from Plato and, with their own additions, imposed this composite system upon us. In this they were greatly aided by the structure of language and psycho-logical habits, which from the primitive down to this very day have affected all of us consciously or unconsciously, and have introduced serious difficulties even in science and in mathematics.

Our rulers: politicians, 'diplomats', bankers, priests of every description, economists, lawyers, etc., and the majority of teachers remain at present largely or entirely ignorant of modern science, scientific methods, structural linguistic and semantic issues of 1933, and they also lack an essential historical and anthropological background, without which a sane orientation is impossible.* This ignorance is often willful as they mostly refuse, with various excuses, to read modern works dealing with such problems. As a result a conflict is created and maintained between the advance of science affecting conditions of actual life and the orientations of our, rulers, which often remain antiquated by centuries, or one or two thousand years. The present world conditions are in chaos; psycho-logically there exists a state of helplessness—hopelessness, often resulting in the feelings of insecurity, bitterness, etc., and we have lately witnessed psychopathological mass outbursts, similar to those of the dark ages. Few of us at present realize that, as long as such ignorance of our rulers prevails, *no solution of our human problems is possible*.

The distinctly novel issue in a non-aristotelian system seems to be that in a human class of life elementary methodological and structural ignorance about the world and ourselves, as revealed by science, is bound to introduce delusional factors, for no one can be free from some conscious or unconscious structural assumptions. The real and only problem therefore seems to be whether our structural assumptions in 1933 are primitive or of the 1933 issue. The older 'popularization of science' is not the solution, it often does harm. The progress of science is due in the main to scientific methods and linguistic revisions, and so the new facts discovered by such methods cannot be properly utilized by antiquated psycho-logical orientations and languages. Such utilization often results only in bewilderment and lack of balance. Before we can adjust ourselves to the new conditions of life, created in the main by science, we must first of all revise our grossly antiquated methods of orientation. Then *only* shall we be able to adjust ourselves properly to the new facts.

Investigations show that the essential scientific structural data of 1933 about the world and ourselves are extremely simple, simpler even than any of the structural fancies of the primitives. We usually have sense enough to fit our shoes to our feet, but not sense enough to revise our older methods of orientation to fit the facts. The elimination of primitive identifications, which is easily accomplished once we take it seriously, produces the necessary psycho-logical change toward sanity.

'Human nature' is not an elementalistic product *of* heredity alone, or of environment alone, but represents a very complex organism-as-a-whole end-result of the enviro-genetic manifold. It seems obvious, once stated, that in-a human class of life, the linguistic, structural, and semantic issues represent powerful and ever present environmental factors, which constitute most important components of all our problems. 'Human nature' *can be changed*, once we know how. Experience and experiments show that this 'change of human nature', which under verbal elementalism was supposed to be impossible, can be accomplished in most cases in a few months, if we attack this problem by the non-elementalistic, *neuro-psycho-logical*, special non-identity technique.

If the ignorance and identifications of our rulers could be eliminated a variety of delusional factors through home and school educational and other powerful agencies would cease to be imposed and enforced upon us, and the revision of our systems would be encouraged, rather than hampered. Effective solutions of our problems would then appear spontaneously and in simple forms; our 'shoes' would fit our 'feet' and we could 'walk through life' in comfort, instead of enduring the present sufferings.

Since our existing systems appear to be in many respects unworkable and involve psychopathological factors owing in the main to certain presuppositions of the aristotelian system, and also for brevity's sake, I call the whole operating systemic complex 'aristotelian'. The outline of a new and modern system built after the rejection of the delusional factors I call 'non-aristotelian'. To avoid misunderstandings I wish to acknowledge explicitly my profound admiration for the extraordinary genius of Aristotle, particularly in consideration of the period in which he lived. Nevertheless, the twisting of his system and the imposed immobility of this twisted system, as enforced for nearly two thousand years by the controlling groups, often under threats of torture and death, have led and can only lead to more disasters. From what we know about Aristotle, there is little doubt that, if alive, he would not tolerate such twistings and artificial immobility of the system usually ascribed to him.

The aim of the work of Aristotle and the work of the non-aristotelians is similar, except for the date of our human development and the advance of science. The problem is whether we shall deal with science and scientific methods of 350 B.C. or of 1941 A.C. In general semantics, in build-ing up a non-aristotelian system, the aims of Aristotle are preserved yet scientific methods are brought up to date.

History shows that the advancement of science and civilization involves, first, an accumulation of observations; second, a preliminary formulation of some kind of 'principles' (which always involve some unconscious assumptions); and, finally, as the numbers of observations increase, it leads to the revision and usually the rejection of unjustified, or false to facts 'principles', which ultimately are found to represent only postulates. Because of the cumulative and non-elementalistic character of human knowledge, a mere challenge to a 'principle' does not carry us far. For expediency, assumptions underlying a system have (1) to be discovered, (2) tested, (3) eventually challenged, (4) eventually rejected, and (5) a *system*, free from the eventually objectionable postulates, has to be built.

Examples of this abound in every field, but the histories of the non-euclidean and non-newtonian systems supply the simplest and most obvious illustrations. For instance, the fifth postulate of Euclid did not satisfy even his contemporaries, but these challenges were ineffective for more than two thousand years. Only in the nineteenth century was the fifth postulate eliminated and non-euclidean systems built without it. The appearance of such systems marked a profound revolution in human orientations. In the twentieth century the much more important 'principles' underlying our notions about the physical world, such as 'absolute simultaneity', 'continuity' of atomic processes, 'certainty' of our experiments and conclusions, etc., were challenged, and systems were then built without them. As a result, we now have the magnificent non-newtonian physics and world outlooks, based on the work of Einstein and the quantum pioneers.

Finally, for the first time in our history, some of the most important 'principles' of all principles, this time in the 'mental world', were challenged by mathematicians. Forr instance the universal validity of the so-called 'logical law of the excluded third' was questioned. Unfortunately; as yet, no full-fledged systems based on this challenge have been formulated, and so it remains largely inoperative, although the possibilities of some non-aristotelian, though elementalistic and unsatisfactory 'logics', are made obvious.

Further researches revealed that the *generality* of the 'law of the excluded third' is not an independent postulate, but that it is only an elementalistic consequence of a deeper, invariably false to facts principle of 'identity', often unconscious and consequently particularly pernicious. Identity is defined as 'absolute sameness in all respects', and it is this 'all' which makes identity impossible. If we eliminate this 'all' from the definition, then the word 'absolute' loses its meaning, we have 'sameness in some respects', but we have no 'identity', and only 'similarity', 'equivalence', 'equality', etc. If we consider that all we deal with represents constantly changing sub-microscopic, interrelated processes which are not; and cannot be 'identical with themselves', the old dictum that 'everything is identical with itself' becomes in 1933 a principle invariably false to facts.

Someone may say, 'Granted, but why fuss so much about it?' My answer would be, 'Identification is found in all known primitive peoples; in all known forms of "mental" ills; and in the great majority of personal, national, and international maladjustments. It is important, therefore, to eliminate such a harmful factor from our prevailing systems.' Certainly no one would care to contaminate his child with a dangerous germ, once it is known that the given factor is dangerous. Furthermore, the results of a complete elimination of identity are so farreaching and beneficial for the daily-life of everyone, and for science, that such 'fussing' is not only justified, but becomes one of the primary tasks before us. Anyone who will study the present work will be easily convinced by observations of human difficulties in life, and science, that the majority of these difficulties arise from necessary false evaluations, in consequence of the unconscious false to facts identifications.

The present work therefore formulates a system, called non-aristotelian, which is based on the complete rejection of identity and its derivatives, and shows what very simple yet powerful structural factors of sanity can be found in science. The experimental development of science and civilization invariably involves more and more refined discriminations. Each refinement means the elimination of some identifications somewhere, but many still remain in a partial and mostly unconscious form. The non-aristotelian system formulates the general problem of non-identity, and gives childishly simple nonelementalistic means for a complete and conscious elimination of identification, and other delusional or psychopathological factors in all known fields of human endeavours, in science, education, and all known phases of private, national, and international life.

The non-aristotelian system presented here has turned out to be a strictly *empirical* science, as predicted, with results which have greatly surpassed even my expectations. General semantics is not any 'philosophy', or 'psychology', or 'logic', in the ordinary sense. It is a new extensional discipline which explains and trains us how to use our nervous systems most efficiently. It is not a medical science, but like bacteriology, it is indispensable for medicine in general, and for psychiatry, mental hygiene, and education in particular. In brief, it is the formulation of a new non-aristotelian system of orientation which affects every branch of science and life. The separate issues involved are not entirely new; their methodological formulation as *a system* which is workable, teachable and so elementary that it can be applied by children, is entirely new.

Perplexities in Theories of 'Meaning'

There is a fundamental confusion between the notion of the older 'semantics' as connected with a theory of *verbal* 'meaning' and words defined by words, and the present theory of 'general semantics' where we deal only with neuro-semantic and *neuro-linguistic living reactions* of Smith$_1$, Smith$_2$, etc., as their reactions to neuro-semantic and neuro-linguistic *environments as environment*.

The present day theories of 'meaning' are extremely confused and difficult, ultimately hopeless, and probably harmful to the sanity of the human race. Of late in the United States some members of the progressive education movement have written much on 'referents' and 'operational' methods, in the abstract, based on verbalism. Let us consider some facts, and how the theories of referents and operational methods fit *human evaluations*. Here is, for instance, Smith$_1$, who, through family, social, economic, political, etc., conditions has become 'insane'. Smith$_1$, finally, in ordinary parlance, kills Smith$_2$. From a human point of view it is a very complex and tragic situation. Let us account for it in terms of referents and operations. The body and the heart of Smith$_2$, the hand of Smith$_1$, the knife, etc., are perfectly good referents. The grabbing of the knife by Smith$_1$ and plunging it in the heart of Smith$_2$, the falling down on the ground by Smith$_2$, and the kicking of his legs are perfectly good operations. However, where *is human evaluation?* Where is concern with 'sanity' and 'insanity'? Here we deal with some of the deepest human and social tragedies which, in this case, involve not only the killing of Smith$_2$ by Smith$_1$, but the sick, unhappy, twisted life of Smith$_1$, affecting all his life connections, and with which we must be concerned if we are to be *human beings* and different from apes.

Such an example is of course extreme and over-simplified, although it illustrates the principles. However, officially teaching such *methods* which are *inadequate* to handle evaluation, and so human values, has a definite sinister effect, among others, on the 'sex' life of the students. Many of them are taught to orient themselves generally by referents and operations only; and so mere physiological performance is often *identified* with mature love life, etc., and is a causative factor in the wide-spread marital unhappiness, promiscuity and other lowerings of. human cultural and ethical standards.

Thus, theories of 'meaning' or still worse, 'meaning of meaning', based on 'referents' and 'operational' methods are thoroughly *inadequate to* account for human values, yet they do affect the nervous systems of humans. We must, therefore, work out *a theory of evaluation* which is based on the optimum electro-colloidal action and reaction of the nervous system.

There is no doubt that a civilized society needs some mature 'morals', 'ethics', etc. In a general theory of evaluation and sanity we must consider seriously such problems, if we are to be sane humans at all. Theory and practice show that healthy, well-balanced people are naturally 'moral' and 'ethical', unless their educations have twisted their types of evaluations. In general semantics we do not 'preach' 'morality' or 'ethics' *as such*, but we train students in consciousness *of* abstracting, consciousness off the multiordinal mechanisms of evaluation, *relational* orientations, etc., which bring about cortico-thalamic integration, and then as a result 'morality', 'ethics', awareness of social responsibilities, etc., follow automatically. Unfortunately our educational systems are unaware of, or even, negativistic toward, such neuro-semantic and neuro-linguistic issues. These are sad observations to be made about our present educational systems.

May I suggest that readers consult *Apes, Men and Morons* and *Why Men Behave Like Apes* by Earnest A. Hooton; *The Mentality of Apes* by W. Kohler, *The Social Life of Apes and Monkeys* by S. Zuckerman, and many other studies of this kind. They might then more clearly understand how the aristotelian type of education leads to the humanly harmful, gross, macroscopic, brutalizing, biological, animalistic types of orientations which are shown today to be *humanly inadequate*. These breed such 'führers' as different Hitlers, Mussolinis, Stalin, etc., whether in political, financial, industrial, scientific, medical, educational, or even publishing, etc., fields, fancying that they represent 'all' of the *human* world ! Such delusions must ultimately be destructive to *human* culture, and responsible for the tragic 'cultural lag', stressed so much today by social anthropologists.

Identifications and Misevaluations

The problem of *general* identification is a major problem which does not seem to be understood at all even by specialists. Psychiatrists know professionally the tragic, consequences of identifications in their patients. But what even psychiatrists do not realize is that identifications in daily life are extremely frequent and bring about every kind of difficulties.

As a matter of fact we live in a world in which non-identity is as entirely general as gravitation, and so *every identification is* bound to be in some degree a mis-evaluation. In a four-dimensional world where 'every geometrical point has a date', even an 'electron' at different dates is not identical with itself, because the sub-microscopic processes actually going on in this world cannot empirically be stopped but only transformed. We can, however, through extensional and four-dimensional methods translate the dynamic into the static and the static into the dynamic, and so establish a similarity of structure between language and facts, which was impossible by aristotelian methods. Unfortunately even some modern physicists are unable to understand these simple facts.

To communicate to my classes what I want to convey to my readers here, the following procedure has been useful. In my seminars I pick a young woman student and pre-arrange with her a demonstration about which the class knows nothing. During the lecture she is called to the platform and I hand her a box of matches which she takes carelessly and drops on the desk. That is the only 'crime' she has committed. Then I begin to call her names, etc., with a display of anger, waving my fists in front of her face, and finally with a big gesture, I slap her face gently. Seeing this 'slap', as a rule ninety per cent of the students recoil and shiver; ten per cent show no overt reactions. The latter have seen what they have seen, but they *delayed their evaluations*. Then I explain to the students that their recoil and shiver was an *organismal evaluation* very harmful in principle, because they *identified the* seen facts with their judgements, creeds, dogmas, etc. Thus their reactions were entirely unjustified, as what they have *seen* turned out to be merely a scientific demonstration of the mechanism of identification, which identification I expected.

Neurological Mechanisms of Extensionalization

There is an especially broad generalization, already referred to, which empirically indicates a fundamental difference between the traditional, aristotelian, intensional orientations, and the new non-aristotelian extensional orientations, and in many ways summarizes the radical differences between the two systems. This is the problem of *intension* (spelled with an s) and *extension*. Aristotle, and his followers even today, recognized the difference between intension and extension. However, they considered the problem *in the abstract*, never applying it to *human living reactions as living reactions*, which can be predominantly intensional or predominantly extensional. The interested reader is advised to consult any textbook on 'logic' concerning 'intension' and 'extension', as well as the material given in this text (see index).

The difference can be illustrated briefly by giving examples of 'defini*tions*'. Thus a 'definition' by intension is given in terms of aristotelian 'properties'. For instance, we may verbally 'define' 'man' as a 'featherless biped', 'rational animal', and what not, which really makes no difference, because no listing of 'properties' could possibly cover 'all' the characteristics of $Smith_1$, $Smith_2$, etc., and their inter-relations.

By extension 'man' is 'defined' by exhibiting a class of individuals made up of $Smith_1$, $Smith_2$, etc.

On the surface this difference may appear unimportant; not so in *living life applications*. The deeper problems of neurological mechanisms enter here. If we orient ourselves predominantly by intension or verbal definitions, our orientations depend mostly on the cortical region. If we orient ourselves by extension or facts, this type of orientation by necessity follows the natural order of evaluation, and involves thalamic factors, introducing automatically cortically delayed reactions. In other words, orientations by intension tend to train our nervous systems in a split between the functions of the cortical and thalamic regions; orientations by extension involve the integration of cortico-thalamic functions.

Orientations by extension induce an *automatic* delay of reactions, which *automatically* stimulates the cortical region and regulates and protects the reactions of the usually over-stimulated thalamic region.

What was said here is elementary from the point of view of neurology. The difficulty is that this little bit of neurological knowledge is not applied in practice. Neurologists, psychiatrists, etc., have treated these problems in an 'abstract', 'academic', detached way only, somehow, entirely unaware that living human reactions depend on the working of the human nervous system, from which dependence there is no escape. No wonder 'philosophers', 'logicians', mathematicians, etc., disregard the working of their nervous systems if even neurologists and psychiatrists still orient themselves by verbal fictions in the '*abstract*'.

If we investigate, it seems appalling how little of the vast knowledge we have is actually applied. Even the ancient Persians showed their understanding of the difference between *learning* and *applying* in their proverb: 'He who learns and learns and yet *does* not what he knows, is one who plows and plows yet never sows'. In this new modern non-aristotelian system we have not only to '*know* elementary facts of modern science, including neuro-linguistic and neuro-semantic researches, but also to *apply* them. In fact, the whole passage from the aristotelian to non-aristotelian systems depends on this change of attitude from intension to extension, from macroscopic to sub-microscopic orientations, from 'objective' to process orientations, from subject-predicate to relational evaluations, etc. This is a laborious process and months of self-discipline are required for adults before these new methods can be applied generally; children as a rule have no difficulties.

If we stop to reflect, however, it seems obvious that those who are trained in two-valued, macroscopic, '*objective*', aristotelian orientations only, are thoroughly unable to have modern, electro-colloidal, sub-microscopic, infinite-valued, *process* orientations in life, which can be acquired only by training in non-aristotelian methods.

It is sad indeed to deal with even young scientists in the colloidal and quantum fields who, after taking off their aprons in the laboratory, relapse immediately into the two-valued, prevalent aristotelian orientations, thus ceasing to be scientific 1941. In many ways these scientists are worse off than the 'man on the street', because of the artificially *accentuated split* between their scientific and their life orientations. Although they work in. an infinite-valued, non-aristotelian field, even they need special training to become conscious of how to apply their own scientific non-aristotelian methods to life problems.

Empirically the consequences of training in the new methods are astonishingly far-reaching. This is easily understood after reflection, because the integrating of the functions of the cortical and thalamic regions brings about better functioning of glands, organs, etc. Although general semantics is not a medical science, we can understand why the non-aristotelian extensional thalamo-cortical methods bring about a great deal of stabilization and even psychosomatic consequences, as the empirical results achieved by my psychiatric co-workers and myself indicate.

Extensional Devices and Some Applications

To achieve extensionalization we utilize what I call 'extensional devices':

1) Indexes	Working Device
2) Dates	Working Device
3) Etc. *(et cetera)*	Safety Device
4) Quotes	Safety Device
5) Hyphens	Safety Device

It should be noticed that in a four-dimensional world dating is only a particular temporal index by which we can deal effectively with spacetime. In non-aristotelian orientations these extensional devices should be used habitually and permanently, with a slight motion of the hands to indicate absolute individuals, events, situations, etc., which change at different dates, also different orders of abstraction, etc. Thus thalamic factors become involved, *without which the coveted thalamo-cortical integration cannot be accomplished.*

I may add that all existing psychotherapy, no matter of what school, is based on the partial and particular extensionalization of a given patient, depending on the good luck and personal skill of the psychiatrist. Unfortunately these specialists are in the main unaware of what is said here, and of the existence of a theory of sanity which gives general, simple, and workable thalamo-cortical methods for extensionalization, and so thalamocortical integration.

A few illustrations of the wide practical applications of some of the devices may be given here. In many instances serious maladjustments follow when 'hate' absorbs the whole of the affective energy of the given individual. In such extreme cases 'hate' exhausts the *limited* affective energy. No energy is left for positive feelings and the picture is often that of a dementia praecox, etc. Thus an individual 'hates' *a generalization* 'mother', 'father', etc., and so by identification 'hates' 'all mothers', 'all fathers', etc., in fact, hates the whole fabric of human society, and becomes a neurotic or even a psychotic. Obviously, it is useless to preach 'love' for those who have hurt and have done the harm. Just the opposite; as a preliminary step, by *indexing* we *allocate* or *limit* the 'hate' to the individual Smith$_1$, instead of a 'hate' for a generalization which spreads over the world. In actual cases we can watch how this allocation or limitation of 'hate' from a generalization to an individual helps the given person. The more they 'hate' the individual Smith$_1$, instead of a generalization, the more positive affective energy is liberated, and the more 'human' and 'normal' they become. It is a long struggle, but so far empirically invariably successful, provided the student is willing to work persistently at himself.

But even this indexed *individualized* 'hate' is not desirable, and we eliminate it rather simply *by dating*. Obviously $Smith_1^{1920}$ is *not* $Smith_1^{1940}$ and most of the time $hurt_1^{1920}$ would not be a 'hurt' in 1940. With such types of orientations the individual becomes adjusted, and serious improvements *in family* and social relationships follow, because the student has trained himself in a general method for handling his own problems.

Similar mechanisms of generalization through identifications are involved in morbid and other generalized *fears* which are so disastrous for *daily* adjustment. Because *thalamic factors are involved*, these difficulties are helped greatly or eliminated by a similar use of the extensional devices to individualize and then date the allocated fears.

What a heavy price we may sometimes pay for the disregard of extensional devices in connection with the structure of language, can be illustrated no better than by the life history and work of Dr. Sigmund Freud. In his writings Freud ascribed *one* intensional undifferentiated 'sex' even to infants, which revolted public opinion. If Freud would have used the extensional devices he would not have gotten into such detrimental professional and other difficulties. He would not have used the fiction 'sex' without indexes, dates and quotes, and he would have explained that an infant has a ticklish organ which could be labelled $sexy_0^0$ at birth, sex_1^{1} at the age of one, sex_2^{2} at the age of two, etc. These are obviously different in life, but the differences are hidden by the one abstract definitional term 'sex', and made obvious only by the extensional techniques. Let us be frank about it. The intensional abstract 'sex' labels a fiction. By extension or facts, 'sex' varies with every individual not only with age (dates), but in relation to endless other factors, and can be handled adequately only by the use of extensional devices.

Implications of the Structure of Language

In what is said above we were already dealing with the change from an intensional to an extensional *structure* of language, and so orientation. We can investigate a step further, and find that the aristotelian structure of language is in the main *elementalistic*, implying, through structure, a split or separation of what in actuality cannot be separated. For instance, we can verbally split 'body' *and* 'mind', 'emotion' *and* 'intellect', 'space' *and* 'time', etc., which as a matter of fact cannot be separated empirically, and can be split only verbally. These elementalistic, splitting, structural characteristics of language have been firmly rooted in us through the aristotelian training. It built for us *a fictitious animistic world* not much more advanced than that of the primitives, a world in which under present conditions an optimum adjustment is in principle impossible,

In a non-aristotelian system we do not use elementalistic terminology to represent facts which are non-elementalistic. We use terms like 'semantic reactions', 'psychosomatic', 'space-time', etc., which eliminate the verbally implied splits, and consequent mis-evaluations. In the beginning of my seminars when I am explaining space-time, students often react by saying, 'Oh, you mean "space" *and* "time"'. This translation would abolish the whole modern advances of physics, because of the structural implications of a delusional verbal split. Similarly the habitual use of the non-elementalistic term 'semantic reactions' eliminates metaphysical and verbal speculations on such elementalistic fictions as 'emotion' *and* 'intellect', etc., considered as separate entities.

Education for Intelligence and Democracy

The new, non-aristotelian types of evaluations are forthcoming in every field of human endeavour, in science and/or life, necessitated by the urgencies of modern conditions. The main problem today is to formulate *general methods by* which these many separate attempts can be unified into a general system of evaluation, which can become communicable to children and, with more difficulty, even to adults. History shows that whenever older methods prove their inefficiency new methods are produced which meet the new conditions more effectively. But the difficulties involved must first be clearly *formulated* before methods and techniques can be devised with which we can deal with them more successfully.

It seems unnecessary to enlarge on the present day world tragedies because many excellent volumes have already been written and are continuing to accumulate, psychiatric evaluations included. I must stress, however, that no writer I know of has ever understood, the depth of the pending transition from the aristotelian system to an already formulated non-aristotelian system. This transition is much deeper than the change from merely one aristotelian 'ism' to another.

We argue so much today about 'democracy' versus 'totalitarianism'. Democracy presupposes intelligence of the masses; totalitarianism does *not to* the same degree. But a 'democracy' without intelligence of the masses under modern conditions can be a worse human mess than any dictatorship could be. Certainly present day education, while it may cram students' heads with some data, without giving them any *adequate methodological synthesis* and extensional working methods, does not train in 'intelligence' and how to become adjusted to life, and so does not work toward 'democracy'. Experiments show that even a root can learn a lesson, and animals can learn by trial and error. But we humans after these millions of years should have learned how to utilize the 'intelligence' which we supposedly have, with some predictability, etc., and use it *constructively*, not *destructively*, as, for example, the Nazis are doing under the guidance of specialists.

In general semantics we believe that some such thing as healthy human intelligence is possible, and so somehow we believe in the eventual possibility of 'democracy'. We work, therefore, at methods which could be embodied even in elementary education to develop the coveted thalamo-cortical integration, and so sane intelligence. Naturally in our work *prevention is* the main aim, and this can be accomplished only through education, and as far as the present is concerned, through re-education, and re-training of the human nervous system.

2. TERMINOLOGY AND MEANINGS

On Semantic Reactions

The term *semantic reaction* is fundamental for the present work and *non-elementalistic systems*. The term 'semantic' is derived from the Greek *semantikos*, 'significant', from *semainein* 'to signify', 'to mean', and was introduced into literature by Michel Breal in his *Essai de Semantique*. The term has been variously used in a more or less general or restricted sense by different writers.

The problems of 'meaning' are very complex and too little investigated, but it seems that 'psychologists' and 'philosophers' are not entirely in sympathy with the attitude of the neurologists. It is necessary to show that in a [non-A]-system, which involves a new theory of meanings based on *non-el* semantics, the neurological attitude toward 'meaning' is the only structurally corrrect and most useful one.

The explanation is quite simple. We start with the negative [non-A] premise that words are *not* the un-speakable objective level, such as the actual objects outside of our skin *and* our personal feelings inside our skin. It follows that the only link between the objective and the verbal world is exclusively structural, necessitating the conclusion that the only content of all 'knowledge' is structural. Now structure can be considered as a complex of relations, and ultimately as multi-dimensional order.

From this point of view, all language can be considered as names either for un-speakable entities on the objective level, be it things or feelings, or as *names for relations*. In fact, even objects, as such, *could* be considered as relations between the sub-microscopic events and the human nervous system. If we enquire what the last relations represent, we find that an object represents an abstraction of low order produced by our nervous system as the result of the sub-microscopic events acting as stimuli upon the nervous system. If the objects represent abstractions of some order, then, obviously, when we come to the enquiry as to language, we find that words are still higher abstractions from objects. Under such conditions, a theory of 'meaning' looms up naturally. If the objects, as well as words, represent abstractions of different order, an individual, A, cannot know what B abstracts, unless B tells him, and so the 'meaning' of a word must be given by a definition. This would lead to the dictionary meanings of words, provided we could define all our words. But this is impossible. If we were to attempt to do so, we should soon find that our vocabulary was exhausted, and we should reach a set of terms which could not be any further defined, from lack of words. We thus see that all linguistic schemes, if analysed far enough, would depend on a set of undefined terms. If we enquire about the 'meaning' of a word, we find that it depends on the 'meaning' of other words used in defining it, and that the eventual new relations posited between them ultimately depend *on the m.o meanings of the undefined terms*, which, at a given period, cannot be elucidated any further.

Naturally, any fundamental theory of 'meaning' cannot avoid this issue, which must be crucial. Here a semantic experiment suggests itself. I have performed this experiment repeatedly on myself and others, invariably with similar results. Imagine that we are engaged in a friendly serious discussion with some one, and that we decide to enquire into the meanings of words. For this special experiment, it is not necessary to be very exacting, as this would enormously and unnecessarily complicate the experiment. It is useful to have a piece of paper and a pencil to keep a record of the progress.

We begin by asking the 'meaning' of every word uttered, being satisfied for this purpose with the roughest definitions; then we ask the 'meaning' of the words used in the definitions, and this process is continued usually for no more than ten to fifteen minutes, until the victim begins to speak in circles—as, for instance, defining 'space' by 'length' and 'length' by 'space'. When this stage is reached, we have come usually to the *undefined* terms of a given individual. If we still press, no matter how gently, for definitions, a most interesting fact occurs. Sooner or later, signs of *affective disturbances* appear. Often the face reddens; there is a bodily restlessness; sweat appears—symptoms quite similar to those seen in a schoolboy who has forgotten his lesson, which he 'knows but cannot tell'. If the partner in the experiment is capable of self-observation, he invariably finds that he feels an internal *affective pressure*, connected, perhaps, with the rush of blood to the brain and probably best expressed in some such words as 'what he "knows" but cannot tell', or the like. Here we have reached the bottom and the foundation of all *non-elementalistic meanings*—the meanings of *undefined terms*, which we 'know' somehow, but cannot tell. In fact, we have reached the un-speakable level. This 'knowledge' is supplied by the lower nerve centres; it represents affective first order effects, and is interwoven and interlocked with other affective states, such as those called 'wishes', 'intentions', 'intuitions', 'evaluation', and many others. It should be noticed that these first order effects have an objective character, as they are un-speakable—are *not* words.

Since 'knowledge', then, is not the first order un-speakable objective level, whether an object, a feeling.; structure, and so relations, becomes the only possible content of 'knowledge' *and of meanings*. On the lowest level of our analysis, when we explore the objective level (the un-speakable feelings in this case), we must try to define every 'meaning' as a conscious feeling of actual, or assumed, or wished., *relations* which pertain to first order objective entities, psycho-logical included, and which can be evaluated by personal, varied, and racial—again un-speakable first order—psychophysiological effects. Because relations can be defined as multi-dimensional order, both of which terms are *non-el*, applying to 'senses' and 'mind', after *naming* the un-speakable entities, all experience can be *described* in terms of relations or multi-dimensional order. The meanings of meanings, in a given case, in a given individual at a given moment, represent composite, affective psycho-logical configurations of all relations pertaining to the case, coloured by past experiences, state of health, mood of the moment, and other contingencies.

If we consistently apply the organism-as-a-whole principle to any psycho-logical analysis, we must conjointly contemplate at least both aspects, the 'emotional' and the 'intellectual', and so *deliberately ascribe* 'emotional' factors to any 'intellectual' manifestation, and 'intellectual' factors to any 'emotional' occurrence. That is why, on human levels, the *el* term 'psychological' must be abolished and a new term *psycho-logical* introduced, in order that we may construct a science.

From what has been said, we see that not only the structure of the world is such that it is made up of absolute individuals, but that meanings in general, and the meanings of meanings in particular—the last representing probably the un-speakable first order effects—also share, in common with ordinary objects, the absolute individuality of the objective level.

The above explains why, by the inherent structure of the world, life, and the human nervous system, human relations are so enormously complex and difficult; and why we should leave no stone unturned to discover beneath the varying phenomena more and more general and invariant foundations on which human understanding and agreement may be based. In mathematics we find the only model in which we can study the invariance of relations under transformations, and hence the need for future psycho-logicians to study mathematics.

It follows from these considerations that any psycho-logical occurrence has a number of aspects, an 'affective', and an 'intellectual', a physiological, a colloidal, and what not. For the science of psychophysiology, resulting in a theory of sanity, the above four aspects are of most importance. As our actual lives are lived on objective, un-speakable levels, and not on verbal levels, it appears, as a problem of evaluation, that the objective level, including, of course, our un-speakable feelings, 'emotions'., is the most important, and that the verbal levels are only auxiliary, sometimes useful, but at present often harmful, because of the disregard of the *s.r.* The role of the auxiliary verbal levels is only fulfilled if these verbal processes are translated back into first order effects. Thus, through verbal intercourse, in the main, scientists discover useful first order abstractions (objective), and by verbal intercourse again, *culture* is built; but this only when the verbal processes affect the unspeakable psycho-logical manifestations, such as our feelings, 'emotions',.

Some extraordinary parrot could be taught to repeat all the verbal 'wisdom' of the world; but, if he survived at all, he would be just a parrot. The repeated noises would not have affected his first order effects—his affects—these noises would 'mean' nothing to him.

Meanings, and the meanings of meanings, with their inseparable affective components, give us, therefore, not only the *non-elementalistic* foundation on which all civilization and culture depends, but a study of the *non-el* mechanisms of meanings, through psychophysiology and general semantics, gives us, also, powerful physiological means to achieve a host of desirable, and to eliminate a large number of undesirable, psycho-logical manifestations.

The physiological mechanism is extremely simple and necessitates a breaking away from the older elementalism. But it is usually very difficult for any given individual to break away from this older elementalism, as it involves the established *s.r*, and to be effective is, by necessity, a little laborious. The working tool of psychophysiology is found in the *semantic reaction*. This can be described as the psycho-logical reaction of a given individual to words and language and other symbols and events in *connection with their meanings*, and the psycho-logical reactions, which *become meanings and relational configurations* the moment the given individual begins to analyse them or somebody else does that for him. It is of great importance to realize that the term 'semantic' is *non-elementalistic*, as it involves conjointly the 'emotional' as well as the 'intellectual' factors.

From the *non-el* point of view, any affect, or impulse, or even human instinct, when made conscious acquires *non-el* meanings, and becomes ultimately a psycho-logical configuration of desirable or undesirable to the individual relations, thus revealing a workable *non-el* mechanism. Psychotherapy, by making the unconscious conscious, and by verbalization, attempts to discover meanings of which the patient was not aware. If the attempt is successful and the individual meanings are revealed, these are usually found to belong to an immature period of evaluation in the patient's life. They are then consciously revised and rejected, and the given patient either improves or is entirely relieved. The condition for a successful treatment seems to be that the *processes should be managed in a non-elementalistic* way. Mere verbal formalism is not enough, because the full *non-elementalistic* meanings to the patient are not divulged; consequently, in such a case, the *s.r* are not affected, and the treatment is a failure.

The *non-el* study of the *s.r* becomes an extremely general scientific discipline. The study of relations, and therefore order, reveals to us the mechanism of *non-el* meanings; and, in the application of an ordinal *physiological* discipline, we gain psychophysiological means by which powerfully to affect, reverse, or even annul, undesirable *s.r.* In psychophysiology we find a *non-el* physiological theory of meanings and sanity.

From the present point of view, all affective and psycho-logical responses to words and other stimuli *involving meanings* are to be considered as *s.r.* What the relation between such responses and a corresponding persistent psycho-logical *state* may be, is at present not clear, although a number of facts of observation seem to suggest that the re-education of the *s.r* results often in a beneficial change in some of these states. But further investigation in this field is needed.

The realization of this difference is important in practice, because most of the psycho-logical manifestations may appear as evoked by some event, and so are to be called responses or reflexes. Such a response, when lasting, should be called a given *state*, perhaps a semantic state, but not a semantic reflex. The term, 'semantic reaction', will be used as covering both semantic reflexes and states. In the present work, we are interested in *s.r*, from a psychophysiological, theoretical and experimental point of view, which include the corresponding states.

If, for instance, a statement or any event evokes some individual's attention, or one train of associations in preference to another, or envy, or anger, or fear, or prejudice., we would have to speak of all such responses on psycho-logical levels as *s.r*. A stimulus was present, and a response followed; so that, by definition, we should speak of a reaction. As the active factor in the stimulus was the individual meanings to the given person, and his response had meanings to him as a first order effect, the reaction must be called a *semantic* reaction.

The present work is written entirely from the *s.r* point of view; and so the treatment of the material, and the language used, imply, in general, a psycho-logical response to a stimulus in connection with meanings, this response being expressed by a number of such words as 'implies', 'follows', 'becomes', 'evokes', 'results', 'feels', 'reacts', 'evaluates', and many others. All data taken from science are selected, and only those which directly enter as factors in *s.r* are given in an elementary outline. The meanings to the individual are dependent, through the influence of the environment, education, languages and their structure, and other factors, on racial meanings called science, which, to a large extent, because of the structural and relational character of science, become physiological semantic factors of the reactions. In fact, science, mathematics, 'logic'., may be considered from a *non-elementalistic* point of view as *generalized* results of *s.r* acceptable to the majority of informed and not heavily pathological individuals.

As a descriptive fact, the present stage of human development is such that with a very few exceptions our nervous systems do not work properly in accordance with their survival structure. In other words, although we have the potentialities for correct functioning in our nervous system, because of the neglect of the physiological control-mechanism of our *s.r*, we have semantic blockages in our reactions, and the more beneficial manifestations are very effectively prevented.

The present analysis divulges a powerful mechanism for the control and education of *s.r*; and, by means of proper evaluation, a great many undesirable manifestations on the psycho-logical level can be very efficiently transformed into highly desirable ones. In dealing with such a fundamental experimental issue as the *s.r*, which have been with us since the dawn of mankind, it is impossible to say new things all the time. Very often the issues involved become 'common sense'; but what is the use, in practice, of this 'common sense', if it is seldom, if ever, applied, and in fact cannot be applied because of the older lack of workable psychophysiological formulations? For instance, what could be simpler or more 'common sense' than the [non-A] premise that an object is *not* words; yet, to my knowledge, no one *fully* applies this, or has *fully acquired* the corresponding *s.r*. Without first acquiring this new *s.r*, it is impossible to discover this error and corresponding *s.r* in others; but as soon as we have trained ourselves, it becomes so obvious that it is impossible to miss it. We shall see, later, that the older *s.r* were due to the lack of structural investigations, to the old structure of language, to the lack of consciousness of abstracting, to the low order conditionality of our conditional reactions (the semantic included), and a long list of other important factors. All scientific discoveries involve *s.r,* and so, once formulated, and the new reactions acquired, the discoveries become 'common sense', and we often wonder why these discoveries were so slow in coming in spite of their 'obviousness'. These explanations are given because they also involve some *s.r;* and we must warn the reader that such evaluations (*s.r*): 'Oh, a platitude!', 'A baby knows that', are very effective *s.r to prevent* the acquisition of the new reactions. This is why the 'discovery of the obvious' is often so difficult; it involves very many of semantic factors of new evaluation and meanings.

A fuller evaluation is only reached at present on racial grounds in two or more generations, and never on individual grounds; which, of course, for *personal generalized adjustment* and happiness, is very detrimental. Similarly, only in the study of racial achievements called science and mathematics can we discover the appropriate *s.r* and the nervous mechanism of these so varied, so flexible, and so fundamental reactions.

In fact, without a structural formulation and a [non-A] revision based on the study of science and mathematics, it is impossible to discover, to control, or to educate these *s.r.* For this reason it was necessary to analyse the semantic factors in connection with brief and elementary considerations taken from modern science. But, when all is said and done, and the important semantic factors discovered, the whole issue becomes extremely simple, and easily applied, even by persons without much education. In fact, because the objective levels are *not* words, the only possible aim of science is to discover *structure,* which, when formulated, is *always simple* and easily understood by everyone, with the exception, of course, of very pathological individuals. We have already seen that structure is to be considered as a configuration of relations, and that relations appear as the essential factors in meanings, and so of *s.r.* The present enquiry, because structural, reveals vital factors of *s.r.* The consequences are extremely simple, yet very important. We see that by a simple *structural re-education* of the *s.r,* which in the great mass of people are still on the level of copying animals in their nervous reactions, we powerfully affect the *s.r,* and so are able to impart very simply, to all, in the most elementary education of the *s.r* of the child, *cultural* results at present sometimes acquired unconsciously and painfully in university education.

It is important to preserve the *non-el* or organism-as-a-whole attitude and terminology throughout, because these represent most important factors in our *s.r.* Sometimes it is necessary to emphasize the origin, or the relative importance, of a given aspect of the impulse or reaction, or to translate for the reader a language not entirely familiar to him into one to which he is more accustomed. In such cases, I use the old *el* terms in quotation marks to indicate that I do not eliminate or disregard the other aspects—a disregard which otherwise would be implied by the use of the old terms.

An important point should be stressed; namely, that the issues are fundamentally simple, because they are similar in structure to the structure of human 'knowledge' and to the nervous structure on which so-called 'human nature' depends. Because of this similarity, it is unconditionally necessary to become fully acquainted with the new terms of new structure, and to use them habitually. then will the beneficial results follow. All languages have some structure; and so all languages involve automatically the, of necessity, interconnected *s.r.* Any one who tries to translate the new language into the old while 'thinking' in the older terms is confronted with an inherent neurological difficulty and involves himself in a hopeless confusion of his own doing. The reader must be warned against making this mistake.

The present system is an interconnected whole: the beginning implies the end, and the end implies the beginning. Because of this characteristic, the book should be read *at least twice*, and preferably oftener. I wish positively to discourage any reader who intends to give it merely a superficial reading.

On the Un-speakable Objective Level

The term 'un-speakable' expresses exactly that which we have up to now practically entirely disregarded; namely, that an object or feeling, say, our toothache, is *not* verbal, is *not* words. Whatever we may say will not be the objective level, which remains fundamentally un-speakable. Thus, we can sit on the object called 'a chair', but we cannot sit on the noise we made or the name we applied to that object. It is of utmost importance for the present [non-A]-system not to confuse the verbal level with the objective level, the more so that all our immediate and direct 'mental' and 'emotional' reactions, and all s.r, states, and reflexes, belong to the un-speakable objective levels, as these are *not* words. This fact is of great, but unrealized, importance for the training of appropriate *s.r*. We can train these reactions simply and effectively by 'silence on the objective levels', using familiar *objects* called 'a chair' or 'a pencil', and this training automatically affects our 'emotions', 'feelings', as well as other psycho-logical immediate responses difficult to reach, which are also *not* words. We can train simply and effectively the *s.r* inside our skins by training on purely objective and familiar grounds outside our skins, avoiding unnecessary psycho-logical difficulties, yet achieving the desired semantic results. The term 'un-speakable' is used in its strict English meaning. The objective level is *not* words, can *not* be reached by words alone, and has nothing to do with 'good' or 'bad'; neither can it be understood as 'non-expressible by words' or 'not to be described by words', because the terms 'expressible' or 'described' already presuppose words and symbols. Something, therefore, which we call 'a chair' or 'a toothache' may be *expressed* or *described* by words; yet, the situation is not altered, because the given description or expression will *not* be the actual objective level which we call 'a chair' or 'a toothache'.

Semantically, this problem is genuinely crucial. Any one who misses that—and it is unfortunately easily missed—will miss one of the most important psycho-logical factors in all *s.r* underlying sanity. This omission is facilitated greatly by the older systems, habits of thought, older *s.r*, and, above all, by the primitive *structure* of our [A] language and the use of the 'is' of identity. Thus, for instance, we *handle* what we call a pencil. Whatever we *handle* is un-speakable; yet we *say* 'this *is* a pencil', which statement is unconditionally false to facts, because the object appears as an absolute individual and *is not* words. Thus our *s.r* are at *once trained in delusional values*, which must be pathological.

I shall never forget a dramatic moment in my experience. I had a very helpful and friendly contact prolonged over a number of years with a very eminent scientist. After many discussions, I asked if some of the special points of my work were clear to him. His answer was, 'Yes, it is all right, and so on, *but*, how can you expect me to follow your work all through, if I still do not know what an object is?' It was a genuine shock to me. The use of the little word 'is' as an identity term applied to the objective level had paralysed most effectively a great deal of hard and prolonged work. Yet, the semantic blockage which prevented him from acquiring the new *s.r* is so simple as to seem trifling, in spite of the semantic harm done. The definite answer may be expressed as follows: 'Say whatever you choose *about* the object, and whatever you might say *is not* it.' Or, in other words: 'Whatever you might *say* the object "is", well it *is not*.' This negative statement is *final*, because it is *negative*.

I have enlarged upon this subject because of its crucial semantic importance. Whoever misses this point is missing one of the most vital factors of practically all *s.r* leading toward sanity. The above is easily verified. In my experience I have never met any one, even among scientists, who would *fully* apply this childish 'wisdom' as an *instinctive* 'feeling' and factor in all his *s.r.* I want also to show the reader the extreme simplicity of a [non-A]-system based on the denial of the 'is' of identity, and to forewarn him against very real difficulties induced by the primitive structure of our language and the *s.r* connected with it. Our actual lives are lived entirely on the objective levels, including the un-speakable 'feelings', 'emotions'., the verbal levels being only *auxiliary*, and effective only if they are translated back into first order un-speakable effects, such as an object, an action, a 'feeling'., all on the silent and un-speakable objective levels. In all cases of which I know at present, where the retraining of our *s.r* has had beneficial effects, the results were obtained when this 'silence on the objective levels' has been attained, which affects all our psycho-logical reactions and regulates them to the benefit of the organism and of his survival adaptation.

On 'copying' in Our Nervous Reactions

The selection of the term 'copying' was forced upon me after much meditation. Its standard meaning implies 'reproduction after a model', applicable even to mechanical processes, and although it does not exclude, it does not necessarily include conscious copying. It is not generally realized to what an important extent copying plays its role in higher animals and man.

Some characteristics are inborn, some are acquired. Long ago, Spalding made experiments with birds. Newly hatched birds were enclosed in small boxes which did not allow them to stretch their wings or to see other birds fly. At the period when usually flying begins, they were released and began to fly at once with great skill, showing that flying in birds is an inborn function. Other experiments were made by Scott to find out if the characteristic song of the oriole was inborn or acquired. When orioles, after being hatched, were kept away from their parents, at a given period they began to sing; but the peculiar melody of their songs was different from the songs of their parents. Thus, singing is an inborn characteristic, but the special melody is due to copying parents, and so is acquired.

In our human reactions, speech in general is an inborn characteristic, but what special language or what special *structure* of language we acquire is due to environment and copying—much too often to unconscious and, therefore, uncritical copying. As to the copying of animals in our nervous reactions, this is quite a simple problem. Self-analysis, which is rather a difficult affair, necessitating a serious and efficient 'mentality', was impossible in the primitive stage. Copying parents in many respects began long before the appearance of man, who has naturally continued this practice until the present day. The results, therefore, are intimately connected with reactions of a pre-human stage, transmitted from generation to generation. But for our present purpose, the most important form of the copying of animals was, and is, the copying of the comparative unconditionality of their conditional reflexes, or lower order conditionality; the animalistic identification or confusion of orders of abstractions, and the lack of consciousness of abstracting, which, while natural, normal, and necessary with animals, becomes a source of endless semantic disturbances for humans. More about copying will be explained as we proceed.

It should be noticed also that because of the structure of the nervous system and the history of its development, the more an organism became 'conscious', the more this copying became a neurological necessity, as exemplified in parrots and apes. With man, owing to the lack of consciousness of abstracting, his copying capacities became also much more pronounced and often harmful. Even the primitive man and the child are 'intelligent' enough to observe and copy, but not informed enough in the racial experiences usually called science, which, for him, are nonexistent, to discriminate between the reactions on the 'psychological' levels of animals and the typical responses which man with his more complex nervous system should have. Only an analysis of *structure* and *semantic reactions*, resulting in consciousness of abstracting, can free us from this unconscious copying of animals, which, let us repeat, must be pathological for man, because it eliminates a most vital regulating factor in human nervous and *s.r*, and so vitiates the whole process. This factor is not simply additive, so that, when it is introduced and *superimposed* on any response of the human nervous system allowing such superimposition, the whole reaction is *fundamentally changed* in a beneficial way.

3. ON STRUCTURE

Any student of science, or of the history of science, can hardly miss two very important tendencies which pervade the work of those who have accomplished most in this field. The first tendency is to base science more and more on experiments; the other is toward greater and more critical verbal rigour. The one tendency is to devise more and better instruments, and train the experimenters; the other is to invent better verbal forms, better forms of representation and of theories, so as to present a more coherent account of the experimental facts.

The second tendency has an importance equal to that of the first; a number of isolated facts does not produce a science any more than a heap of bricks produces a house. The isolated facts must be put in order and brought into mutual structural relations in the form of some theory. Then, only, do we have a science, something to start from, to analyse, ponder on, criticize, and improve. Before this something can be criticized and *improved*, it must first be produced, so the investigator who discovers some fact, or who formulates some scientific theory, does not often waste his time. Even his errors may be useful, because they may stimulate other scientists to investigate and improve.

Scientists found long ago that the common language in daily use is of little value in science. This language gives us a form of representation of very old structure in which we find it impossible to give a full, coherent account of ourselves or of the world around us. Each science has to build a special terminology adapted to its own special purposes. This problem of a suitable language is of serious importance. Too little do we realize what a hindrance a language of antiquated structure is. Such a language does not help, but actually prevents, correct analysis through the semantic habits and structural implications embodied in it. The last may be of great antiquity and bound up, by necessity, with primitive-made structural implications, or, as we say, metaphysics, involving primitive *s.r.*

The above explains why the popularization of science is such a difficult and, perhaps, even a semantically dangerous problem. We attempt to translate a creative and correct language which has a structure similar to the structure of the experimental facts into a language of different structure, entirely foreign to the world around us and ourselves. Although the popularization of science will probably remain an impossible task, it remains desirable that the *results* of science should be made accessible to the layman, if means could be found which do not, by necessity, involve misleading accounts. It seems that such methods are at hand and these involve *structural* and semantic considerations.

The term 'structure' is frequently used in modern scientific literature, but, to the best of my knowledge, only Bertrand Russell and Wittgenstein have devoted serious attention to this problem, and much remains to be done. These two authors have analysed or spoken about the structure of propositions, but similar notions can be generalized to languages considered as-a-whole. To be able to consider the structure of one language of a definite structure, we must produce another language of a *different* structure in which the structure of the first can be analysed. This procedure seems to be new when actually performed, although it has been foreseen by Russell. If we produce a [non-A]-system based on 'relations', 'order', 'structure'., we shall be able to discuss profitably the [A]-system which does not allow asymmetrical relations, and so cannot be analysed by [A] means.

The dictionary meaning of 'structure' is given somewhat as follows; Structure: Manner in which a building or organism or other complete whole is constructed, supporting framework or whole of the essential parts of' something (the structure of a house, machine, animal, organ, poem, sentence; sentence of loose structure, its structure is ingenious; ornament should emphasize and not disguise the lines of structure)'. The implications of the term 'structure' are clear, even from its daily sense. To have 'structure' we must have a complex of ordered and interrelated parts.

One of the fundamental functions of 'mental' processes is to distinguish. We distinguish objects by certain characteristics, which are usually expressed by adjectives. If, by a higher order abstraction, we consider individual objects, not in some perfectly *fictitious* 'isolation', but as they appear empirically, as members of some aggregate or collection of objects, we find characteristics which belong to the collection and not to an 'isolated' object. Such characteristics as arise from the fact that the object belongs to a collection are called 'relations'.

In such collections, we have the possibility of *ordering* the objects, and so, for instance, we may discover a relation that one object is 'before' or 'after' the other, or that A is the father of B. There are many ways in which we can order a collection, and many relations which we can find. It is important to notice that 'order' and 'relations' are, for the most part, empirically present and that, therefore, this language is fit to represent the facts as we know them. The structure of the actual world is such that it is *impossible* entirely to isolate an object. An A subject-predicate language, with its tendency to treat objects as in isolation and to have no place for relations (impossible in complete 'isolation') , obviously has a structure not similar to the structure of the world, in which we deal *only* with collections, of which the members are related.

Obviously, under such empirical conditions, only a language originating in the analysis of collections, and, therefore, a language of 'relations', 'order'., would have a *similar structure* to the world around us. From the use of a subject-predicate form of language alone, many of our fallacious anti-social and 'individualistic' metaphysics and *s.r* follow, which we will not analyse here, except to mention that their structural implications follow the structure of the language they use.

If we carry the analysis a step further, we can find relations between relations, as, for instance; the *similarity of relations*. We follow the definition of Russell. Two relations are said to be similar if there is a *one-one* correspondence between the terms of their fields, which is such that, whenever two terms have the relation P, their correlates have the relation Q. and vice versa. For example, two series are similar when their terms can be correlated without change of order, an accurate map is similar to the territory it represents, a book spelt phonetically is similar to the sounds when read,.

When two relations are similar, we say that they have a *similar structure*, which is defined as the class of all relations similar to the given relation.

We see that the terms 'collection', 'aggregate', 'class', 'order', 'relations', 'structure' are interconnected, each implying the others. If we decide to face empirical 'reality' boldly, we must accept the Einstein-Minkowski four-dimensional language, for 'space' and 'time' *cannot be separated empirically*, and so we must have a language of *similar structure* and consider the facts of the world as series of *interrelated ordered events*, to which, as above explained we must ascribe 'structure'. Einstein's theory, in contrast to Newton's theory, gives us such a *language, similar in structure* to the empirical facts as revealed by science 1933 *and common experience*.

The above definitions are not entirely satisfactory for our purpose. To begin with, let us give an illustration, and indicate in what direction some reformulation could be made.

Let us take some actual territory in which cities appear in the following order: Paris, Dresden, Warsaw, when taken from the West to the East. If we were to build a *map* of this territory and place Paris *between* Dresden and Warsaw thus:

Actual Territory	Paris --------Dresden--------Warsaw
Map	Dresden-----Paris------------Warsaw

We should say that the map was wrong, or that it was an incorrect map, or that the map has a *different structure* from the territory. If, speaking roughly, we should try, in our travels, to orient ourselves by such a map, we should find it misleading. It would lead us astray, and we might waste a great deal of unnecessary effort. In some cases, even, a map of wrong structure would bring actual suffering and disaster, as, for instance, in a war, or in the case of an urgent call for a physician.

Two important characteristics of maps should be noticed. A map *is not* the territory it represents, but, if correct, it has a *similar structure* to the territory, which accounts for its usefulness. If the map could be ideally correct, it would include, in a reduced scale, the map of the map; the map of the map, of the map; and so on, endlessly, a fact first noticed by Royce.

If we reflect upon our languages, we find that at best they must be considered *only as maps*. A word *is not* the object it represents; and languages exhibit also this peculiar self-reflexiveness, that we can analyse languages by linguistic means. This self-reflexiveness of languages introduces serious complexities, which can only be solved by the theory of multiordinality, given in Part VII. The disregard of these complexities is tragically disastrous in daily life and science.

It has been mentioned already that the known definitions of structure are not entirely satisfactory. The terms 'relation', 'order', 'structure' are interconnected by implication. At present, we usually consider order as a kind of relation. With the new four-dimensional notions taken from mathematics and physics, it may be possible to treat relations and structure as a form of *multi-dimensional order*. Perhaps, theoretically, such a change is not so important, but, from a practical, applied, educational, and semantic point of view, it seems very vital. Order seems *neurologically simpler* and more fundamental than relation. It is a characteristic of the empirical which we recognize directly by our lower nervous centres ('senses'), and with which we can deal with great accuracy by our higher nervous centres ('thinking'). This term seems most distinctly of the organism-as-a-whole character, applicable both to the activities of the higher, as well as lower, nervous centres, and so *structurally* it must be fundamental.

The rest of this volume is devoted to showing that the common, [A]-system and language which we inherited from our primitive ancestors *differ entirely in structure* from the well-known and established 1933 structure of the world, ourselves and our nervous systems included. Such antiquated map-language, by necessity, must lead us to semantic disasters, as it imposes and reflects its *unnatural* structure on the structure of our doctrines and institutions. Obviously, under such *linguistic* conditions, a science of man was impossible; differing in structure from our nervous system, such language must also disorganize the functioning of the latter and lead us away from sanity.

This once understood, we shall see clearly that researches into the structure of language and the adjustment of this structure to the structure of the world and ourselves, as given by science at each date, must lead to new languages, new doctrines, institutions., and, in fine, may result in a new and saner civilization, involving new s.r which may be called the scientific era.

The introduction of a few new, and the rejection of some old, terms suggests desirable structural changes, and adjusts the structure of the language-map to the known structure of the world, ourselves, and the nervous system, and so leads us to new s.r and a theory of sanity.

As words *are not* the objects which they represent, *structure, and structure alone*, becomes the only link which connects our verbal processes with the empirical data. To achieve adjustment and sanity and the conditions which follow from them, we must study structural characteristics of this world *first,* and, then only, build languages of similar structure, instead of habitually ascribing to the world the primitive structure of our language. All our doctrines, institutions., depend on verbal arguments. If these arguments are conducted in a language of wrong and unnatural structure, our doctrines and institutions must reflect that linguistic structure and so become unnatural, and inevitably lead to disasters.

That languages, as such, all have some structure or other is a new and, perhaps, unexpected notion. Moreover, every language having a structure, by the very nature of language, reflects in its own structure that of the world as assumed by those who evolved the language. In other words, we read unconsciously into the world the structure of the language we use. The guessing and ascribing of a fanciful, mostly primitive-assumed, structure to the world is precisely what 'philosophy' and 'metaphysics' do. The empirical search for world-structure and the building of new languages (theories), of necessary, or similar, structure, is, on the contrary, what science does. Any one who will reflect upon these structural peculiarities of language cannot miss the semantic point that the scientific method uses the only correct language-method. It develops in the *natural order,* while metaphysics of every description uses the reversed, and ultimately a pathological, order.

Since Einstein and the newer quantum mechanics, it has become increasingly evident that the only content of 'knowing' is of a *structural* character; and the present theory attempts a formulation of this fact in a generalized way. If we build a [non-A]-system by the aid of new terms and of methods excluded by the [A]-system, and stop some of our primitive habits of 'thought' and *s.r,* as, for instance, the confusion of order of abstractions, reverse the reversed order, and so introduce the natural order in our analysis, we shall then find that all human 'knowing' exhibits a structure similar to scientific knowledge, and appears as the *'knowing' of structure.* But, in order to arrive at these results, we must depart completely from the older systems, and must abandon permanently the use of the 'is' of identity.

It would seem that the overwhelming importance *for mankind* of systems based on 'relations', 'order', 'structure'., depends on the fact that such terms allow of an exact and 'logical' treatment, as two relations of similar structure have all their logical characteristics in common. It becomes obvious that, as in the [A]-system we could not deal in such terms, higher rationality and adjustment were impossible. It is not the human 'mind' and its 'finiteness' which is to be blamed, but a primitive language, with a structure foreign to this world, which has wrought havoc with our doctrines and institutions.

The use of the term 'structure' does not represent special difficulties when once we understand its origin and its meanings. The main difficulty is found in the old [A] habits of speech, which do not allow the use of structure, as, indeed, this notion has no place in a complete [A] subject-predicativism.

Let us repeat once more the two crucial *negative* premises as established firmly by *all* human experience: (1) Words *are not* the things we are speaking about; and (2) There *is no* such thing as an object in absolute isolation.

These two most important *negative* statements cannot be denied. If any one chooses to deny them, the burden of the proof falls on him. He has to establish what he affirms, which is obviously impossible. We see that it is safe to start with such solid *negative* premises, translate them into positive language, and build a [non-A]-system.

If words *are not* things, or maps *are not* the actual territory, then, obviously, the only possible link between the objective world and the linguistic world is found in *structure, and structure alone*. The only usefulness of a map or a language depends on the *similarity of structure* between the empirical world and the map-languages. If the structure is not similar, then the traveller or speaker is led astray, which, in serious human life-problems, must become always eminently harmful, . If the structures *are similar*, then the empirical world becomes 'rational' to a potentially rational being, which means no more than that verbal, or map-predicted characteristics, which follow up the linguistic or map-structure, are applicable to the empirical world.

In fact, in structure we find the mystery of rationality, adjustment., and we find that the whole content of knowledge is exclusively structural. If we want to be rational and to understand anything at all, we must look for structure, relations, and, ultimately, multi-dimensional order, all of which was impossible in a broader sense in the [A]-system, as will be explained later on.

Having come to such important *positive* results, starting with undeniable *negative* premises, it is interesting to investigate whether these results are *always* possible, or if there are limitations. The second *negative* premise; namely, that there *is no* such thing as an object in absolute isolation, gives us the answer. If there is no such thing as an absolutely isolated object, then, at least, we have two objects, and we shall *always* discover some relation between them, depending on our interest, ingenuity, and what not. Obviously, for a man to speak about anything at all, *always* presupposes *two* objects at least; namely, the object spoken about and the speaker, and so a *relation* between the two is always present. Even in delusions, illusions, and hallucinations, the situation is not changed; because our immediate feelings are also un-speakable and *not* words.

The semantic importance of the above should not be minimized. If we deal with organisms which possess an inherent activity, such as eating, breathing., and if we should *attempt to build for them conditions* where such activity would be impossible or hampered, these *imposed* conditions would lead to degeneration or death.

Similarly with 'rationality'. Once we find in this world at least potentially rational organisms, we should not *impose* on them conditions which hamper or prevent the exercise of such an important and inherent function. The present analysis shows that, under the all-pervading aristotelianism in daily life, asymmetrical relations, and thus structure and order, have been impossible, and so we have been *linguistically* prevented from supplying the potentially 'rational' being with the means for rationality. This resulted in a semi-human so-called 'civilization', based on our copying animals in our nervous process, which, by necessity, involves us in arrested development or regression, and, in general, disturbances of some sort.

Under such conditions, which, after all, may be considered as firmly established, because this investigation is based on undeniable *negative* premises, there is no way out but to carry the analysis through, and to build up a [non-A]-system based on *negative* fundamental premises or the denial of the 'is' of identity with which rationality will be possible.

Perhaps an illustration will make it clearer, the more that the old subject-predicate language rather conceals structure. If we take a statement, 'This blade of grass is green', and analyse it only as a statement, superficially, we can hardly see how any structure could be implied by it. This statement may be analysed into substantives, adjectives, verb.; yet this would not say much about its structure. But if we notice that these words can also make a question, 'Is this blade of grass green?', we begin to realize that the *order* of the words plays an important role in some languages connected with the meanings, and so we can immediately speak of the structure of the sentence. Further analysis would disclose that the sentence under consideration has the subject-predicate form or structure.

If we went to the objective, silent, un-speakable level, and analysed this objective blade of grass, we should discover various structural characteristics in the blade; but these are not involved in the statement under consideration, and it would be illegitimate to speak about them. However, we can carry our analysis in another direction. If we carry it far enough, we shall discover a very intricate, yet definite, relation or complex of relations between the objective blade of grass and the observer. Rays of light impinge upon the blade, are reflected from it, fall on the retina of our eye, and produce within our skins the feeling of 'green'., an extremely complex process which has some definite structure. We see, thus, that any statement referring to anything objective in this world can always be analysed into terms of relations and structure, and that it involves also definite structural assumptions. More than that, as the only possible content of knowledge and science is structural, whether we like it or not, to *know* anything we must search for structure, or posit some structure. Every statement can also be analysed until we come to definite structural issues. This applies, however, with certainty only to significant statements, and, perhaps, not to the various noises which we can make with our mouth with the semblance of words, but which are meaningless, as they are not symbols for anything. It must be added that in the older systems we did not discriminate between words (symbols) and noises (not symbols). In a [non-A]-system such a discrimination is essential.

The structure of the world is, in principle, *unknown;* and the only aim of knowledge and science is to discover this structure. The structure of languages is potentially *known,* if we pay attention to it. Our only possible procedure in advancing our knowledge is to match our verbal structures, often called theories, with empirical structures, and see if our verbal predictions are fulfilled empirically or not, thus indicating that the two structures are either similar or dissimilar.

We see, thus, that in the investigation of structure we find not only means for rationality and for adjustment, and so sanity, but also a most important tool for exploring this world and scientific advance.

From the educational point of view, also, the results of such an investigation seem to be unusually important, because they are extremely simple, *automatic* in their working, and can be applied universally in elementary education. As the issue is merely one of linguistic structure, it is enough to train children to abandon the 'is' of identity, in the habitual use of a *few new terms,* and to warn them repeatedly against the use of some terms of antiquated structure. We shall thus eliminate the pre-human and primitive semantic factors included in the structure of a primitive language. The moralizing and combating of primitive-made metaphysics is not effectual; but the habitual use of a language of modern structure, free from identity, produces semantic results where the old failed. Let us repeat again, a most important point, that the new desirable semantic results follow as *automatically* as the old undesirable ones followed.

In the following enquiry an attempt to build a science of man, or a *non-aristotelian system,* or a theory of sanity, is made, and it will be necessary to introduce a few terms of new structure and to abide by them.

4. GENERAL LINGUISTIC AND SYMBOLISM

In speaking of linguistic researches, I do not mean only an analysis of printed 'canned chatter', as Clarence Day would call it, but I mean the behaviour, the performance, *s.r* of living Smiths and Browns and the connections between the noises uttered by them and their behaviour. No satisfactory analysis has been made, and the reason seems to be in the fact that each existing language really represents a conglomeration of *different* languages with different structures and is, therefore, extremely complex as long as structure is disregarded. That 'linguists', 'psychologists', 'logicians'., were, and usually are, very innocent of *mathematics*, a type of language of the greatest simplicity and perfection, with a clear-cut structure, similar to the structure of the world, Seems to be responsible for this helplessness. Without a study of mathematics, the adjustment of structure seems impossible.

We should not be surprised to find that mathematics must be considered a language. By definition, whatever has symbols and propositions is called a language, a form of representation for this something-going-on which we call the world and which is admittedly *not words*. Several interesting statements can be made about mathematics considered as a language. First of all, mathematics appears as a form of human behaviour, as genuine a human activity as eating or walking, a function in which the human nervous system plays a very serious part. Second, from an empirical point of view a curious question arises: why, of all forms of human behaviour, has mathematizing proved to be *at each historical period* the most excellent human activity, producing results of such enormous importance and unexpected validity as not to be comparable with any other musings of man? Briefly, it may be said that the secret of this importance and the validity of mathematics lie in the mathematical *method* and structure, which the mathematizing Smith, Brown, and Jones have used—we may even say, were *forced* to use. It is not necessary to assume that the mathematicians were 'superior' men. We will see later that mathematics is not a very superior activity of the 'human mind', but it is perhaps the *easiest*, or simplest activity; and, therefore, it has been possible to produce a structurally perfect product of this simple kind.

The understanding and proper evaluation of what has been said about the structure and method of mathematics will play a serious semantic role all through this work, and, therefore, it becomes necessary to enlarge upon the subject. We shall have to divide the abstractions we make into two classes: (1) objective or physical abstractions, which include our daily-life notions; and (2) mathematical abstractions, at present taken from pure mathematics, in a restricted sense, and later generalized. As an example of a mathematical abstraction, we may take a mathematical circle. A circle is defined as the locus of all points in a plane at equal distance from a point called the centre. If we enquire whether or not there is such an actual thing as a circle, some readers may be surprised to find that a mathematical circle must be considered a pure fiction, having nowhere any objective existence. In our definition of a mathematical circle, *all particulars* were included, and whatever we may find about this mathematical circle later on will be strictly dependent on this definition, and no new characteristics, not already included in the definition, will ever appear. We see, here, that *mathematical abstractions are characterized by the fact that they have all particulars included.*

If, on the other hand, we draw an objective 'circle' on a blackboard or on a piece of paper, simple reflection will show that what we have drawn is not a mathematical circle, but a *ring*. It has colour, temperature, thickness of our chalk or pencil mark, . When we draw a 'circle', it is no longer a mathematical circle with *all particulars included in the definition*, but it becomes a physical *ring* in which *new characteristics* appear not listed in our definition.

From the above observations, very important consequences follow. Mathematizing represents a very simple and easy human activity, because it deals with fictitious entities with all particulars included, and we proceed by remembering. The structure of mathematics, because of this over-simplicity, yet structural similarity with the external world, makes it possible for man to build verbal systems of remarkable validity.

Physical or daily-life abstractions differ considerably from mathematical abstractions. Let us take any actual object; for instance, what we call a pencil. Now, we may describe or 'define' a 'pencil' in as great detail as we please, yet it is impossible to include all the characteristics which we may discover in this actual objective pencil. If the reader will try to give a 'complete' description or a 'perfect' definition of any actual physical object, so as to include 'all' particulars, he will be convinced that this task is humanly impossible. These would have to describe, not only the numerous rough, macroscopic characteristics, but also the microscopic details, the chemical composition and changes, sub-microscopic characteristics and the endlessly changing relationship of this objective something which we have called pencil to the rest of the universe., an inexhaustible array of characteristics which could never be terminated. In general, physical abstractions, including daily-life abstractions are such that *particulars are left out*—we proceed by a process of forgetting. In other words, no description or 'definition' will ever include all particulars.

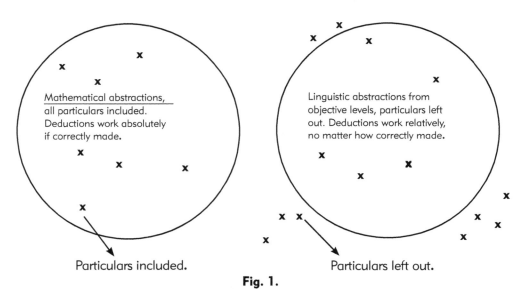

Mathematical abstractions, all particulars included. Deductions work absolutely if correctly made.

Linguistic abstractions from objective levels, particulars left out. Deductions work relatively, no matter how correctly made.

Particulars included.　　　　Particulars left out.

Fig. 1.

Only and exclusively in mathematics does deduction, if correct, work absolutely, for no particulars are left out which may later be discovered and force us to modify our deductions.

Not so in abstracting from physical objects. Here, particulars are left out; we proceed by forgetting, our deductions work only relatively, and must be revised continuously whenever new particulars are discovered. In mathematics, however, we build for ourselves a fictitious and *over-simplified* verbal world, with abstractions which have all particulars included. If we compare mathematics, taken as a language, with our daily language, we see readily that in both verbal activities we are building for ourselves forms of representation for this something-going-on, which is *not* words.

Considered as a language, mathematics appears as a language of the highest perfection, but at its lowest development. Perfect, because the structure of mathematics makes it possible to be so (all characteristics included, and no physical content), and because it is a language of *relations* which are also found in this world. At the lowest development, because we can speak in it as yet about very little and that in a very narrow, restricted field, and with limited aspects.

Our other languages would appear, then, as the other extreme, as the highest mathematics, but also at their lowest development—highest mathematics, because in them we can speak about everything; at their lowest development because they are still [A] and not based on asymmetrical relations. Between the two languages there exists as yet a large unbridged structural gap. The bridging of this gap is the problem of the workers of the future. Some will work in the direction of inventing new mathematical methods and systems, bringing mathematics closer in scope and adaptability to ordinary language (for instance, the tensor calculus, the theory of groups, the theory of sets, the algebra of states and observables, .). Others will undertake linguistic researches designed to bring ordinary language closer to mathematics (for instance, the present work). When the two forms of representation meet on relational grounds, we shall probably have a simple language of mathematical structure, and mathematics, as such, might then even become obsolete.

From what has now been said, it is probably already obvious that if any one wants to work scientifically on problems of such enormous complexity that they have so far defied analysis, he would be helped enormously if he would train his *s.r* in the simplest forms of correct 'thinking'; that is, become acquainted with mathematical methods. The continued application of this relational method should finally throw some light on the greatest complexities, such as life and man. In contrast to enormous advances in all technical fields, our knowledge of 'human nature' has advanced very little beyond what primitives knew about themselves. We have tried to analyse the most baffling phenomena while disregarding structural peculiarities of languages and thus failing to provide sufficient fundamental training in new *s.r.* In practically all universities at present, the mathematical requirements, even for scientists, are extremely low, much lower, indeed, than is necessary for the progress of these scientists themselves. Only those who specialize in mathematics receive an advanced training, but, even with them, little attention is devoted to *method* and *structure* of languages *as such*. Until lately, mathematicians themselves were not without responsibility for this. They treated mathematics as some kind of 'eternal verity', and made a sort of religion out of it; forgetting, or not knowing, that these 'eternal verities' last only so long as the nervous systems of Smiths and Browns are not altered. Besides, many, even now, disclaim any possible connection between mathematics and human affairs. Some of them seem, indeed, in their religious zeal, to try to make their subjects as difficult, unattractive, and mysterious as possible, to overawe the student. Fortunately, a strong reaction against such an attitude is beginning to take place among the members of the younger mathematical generation. This is a very hopeful sign, as there is little doubt that, without the help of professional mathematicians who will understand the general importance of *structure* and *mathematical methods*, we shall not be able to solve our human problems in time to prevent quite serious break-downs, since these solutions ultimately depend on structural and semantic considerations.

The moment we abandon the older theological attitude toward mathematics, and summon the courage to consider it as a form of human behaviour and the expression of *generalized s.r*, some quite interesting problems loom up. Terms like 'logic' or 'psychology' are applied in many different senses, but, among others, they are used as labels for certain disciplines called sciences. 'Logic' is defined as the 'science of the laws of thought'. Obviously, then, to produce 'logic' we should have to study *all* forms of human behaviour connected directly with mentation; we should have to study not only the mentations in the daily life of the average Smiths, Browns., but we should have to study the mentations of Joneses and Whites when they use their 'mind' at its best;

namely, when they mathematize, scientize., and we should also have to study the mentations of those whom we call 'insane', when they use their 'mind' at its worst. It is not our aim to give a detailed list of these forms of human behaviour which we should study, since all should be studied. It is enough for our purpose to emphasize the two main omissions; namely, the study of mathematics and the study of 'insanity'.

As a similar reasoning applies to 'psychology', we must sadly admit that we have as yet no general theory which deserves the name of 'logic' or psycho-logics. What has passed under the name of 'logic', for instance, is not 'logic' according to its own definition, but represents a philosophical grammar of a primitive-made language, of a structure different from the structure of the world, unfit for serious use. If we try to apply the rules of the old 'logic', we find ourselves blocked by ridiculous impasses. So, naturally, we discover that we have no use for such a 'logic'.

It follows also that any one who has any serious intention of becoming a 'logician' or a psycho-logician must, first of all, be a thorough mathematician and must also study 'insanity'. Only with such preparation is there any possibility of becoming a psycho-logician or semantician. Sometimes it is useful to stop deceiving ourselves; and it is deceiving ourselves if we claim to be studying *human* psycho-logics, or *human* 'logic', when we are generalizing only from those forms of human behaviour which we have in common with the animals and neglect other forms, especially the most characteristic forms of human behaviour, such as mathematics, science, and 'insanity'. If, as psycho-logicians, we want to be 'behaviourists', it is clear that we must study *all* known forms of human behaviour. But it seems never to have occurred to the 'behaviourists' that mathematics and 'insanity' are very characteristic forms of human behaviour.

Here, perhaps, it may be advisable to interpolate a short explanation. When we deal with human affairs and man, we sometimes use the term 'ought', which is very often used arbitrarily, dogmatically, and absolutistically, and so its use has become discredited. In many quarters, this term is very unpopular, and, it must be admitted, justly so. My use of it is that of the engineer, who undertakes to study a machine entirely unknown to him—let us say, a motorcycle. He would study and analyse *its structure*, and, finally, would give a verdict that with such a structure, under certain circumstances, this machine *ought* to work in a particular way.

In the present volume, this engineering attitude is preserved. We shall investigate the structure of human knowledge, and we shall conclude that with such a structure it should work in this particular way. In the motorcycle example, the proof of the correctness of the reasoning of the engineer would be to fill the tank with gasoline and make the motorcycle go. In our analogous task, we have to *apply* the information we get and see if it works. In the experiments mentioned above, the [non-A]-system actually has worked, and so there is some hope that it is correct. Further investigations will, of course, add to, or modify, the details, but this is true of all theories.

The affairs of man are conducted by our own, man-made rules and according to man-made theories. Man's achievements rest upon the use of symbols. For this reason, we must consider ourselves as a symbolic, semantic class of life, and those who rule the symbols, rule us. Now, the term 'symbol' applies to a variety of things, words and money included. A piece of paper, called a dollar or a pound, has very little value if the other fellow refuses to take it; so we see that money must be considered as a symbol for human agreement, as well as deeds to property, stocks, bonds, . The *reality* behind the money-symbol is doctrinal, 'mental', and one of the most precious characteristics of mankind. But it must be used properly; that is, with the proper understanding of its structure and ways of functioning. It constitutes a grave danger when misused.

When we say 'our rulers', we mean those who are engaged in the manipulation of symbols. There is no escape from the fact that they do, and that they always will, rule mankind, because we constitute a symbolic class of life, and we cannot cease from being so, except by regressing to the animal level.

The hope for the future consists in the understanding of this fact; namely, that we shall always be ruled by those who rule symbols, which will lead to scientific researches in the field of symbolism and *s.r.* We would then *demand* that our rulers should be *enlightened* and *carefully selected*. Paradoxical as it may seem, such researches as the present work attempts, will ultimately do more for the stabilization of human affairs than legions of policemen with machine guns, and bombs, and jails, and asylums for the maladjusted.

A complete list of our rulers is difficult to give; yet, a few classes of them are quite obvious. Bankers, priests, lawyers and politicians constitute one class and work together. They do not *produce* any values, but manipulate values produced by others, and often pass signs for no values at all. Scientists and teachers also comprise a ruling class. They produce the main values mankind has, but, at present, they do not realize this. They are, in the main, themselves ruled by the cunning methods of the first class.

In this analysis the 'philosophers' have been omitted. This is because they require a special treatment. As an historical fact, many 'philosophers' have played an important and, to be frank, sinister role in history. At the bottom of any historical trend, we find a certain 'philosophy', a structural implication cleverly formulated by some 'philosopher'. The reader of this work will later find that most 'philosophers' gamble on multiordinal and *el* terms, which have *no definite single (one-valued) meaning*, and so, by cleverness in twisting, can be made to appear to mean anything desired. It is now no mystery that some quite influential 'philosophers' were 'mentally' ill. Some 'mentally' ill persons are tremendously clever in the manipulation of words and can sometimes deceive even trained specialists. Among the clever concoctions which appear in history as 'philosophic' systems, we can find flatly opposing doctrines. Therefore, it has not been difficult at any period for the rulers to select a cleverly constructed doctrine perfectly fitting the ends they desired.

One of the main characteristics of such 'philosophers' is found in the delusion of grandeur, the 'Jehovah complex'. Their problems have appeared to them to be above criticism or assistance by other human beings, and the correct procedure known only to super-men like themselves. So quite naturally they have usually refused to make enquiries. They have refused even to be informed about scientific researches carried on outside the realms of their 'philosophy'. Because of this ignorance, they have, in the main, not even suspected the importance of the problems of structure.

In all fairness, it must be said that not all so-called 'philosophy' represents an episode of semantic illness, and that a few 'philosophers' really do important work. This applies to the so-called 'critical philosophy' and to the *theory of knowledge* or *epistemology*. This class of workers I call epistemologists, to avoid the disagreeable implications of the term 'philosopher'. Unfortunately, epistemological researches are most difficult, owing mainly to the lack of scientific psycho-logics, general semantics, and investigations of structure and *s.r.* We find only a very few men doing this work, which, in the main, is still little known and unapplied. It must be granted that their works do not make easy reading. They do not command headlines; nor are they aided and stimulated by public interest and help.

It must be emphasized again that as long as we remain humans (which means a symbolic class of life), the rulers of symbols will rule us, and that no amount of revolution will ever change this. But what mankind has a right to ask—and the sooner the better—is that our rulers should not be so shamelessly ignorant and, therefore, pathological in their reactions. If a psychiatrical and scientific enquiry were to be made upon our rulers, mankind would be appalled at the disclosures.

We have been speaking about 'symbols', but we have not yet discovered any general theory concerning symbols and symbolism. Usually, we take terms lightly and never 'think' what kind of implication and *s.r* one single important term may involve. 'Symbol' is one of those important terms, weighty in meanings. If we use the term 'food', for instance, the presupposition is that we take for granted the existence of living beings able to eat; and, similarly, the term 'symbol' implies the existence of intelligent beings. The solution of the problem of symbolism, therefore, presupposes the solution of the problem of 'intelligence' and structure. So, we see that the issues are not only serious and difficult, but, also, that we must investigate a semantic field in which very little has been done.

In the rough, a symbol is defined as a sign which stands for something. Any sign is not necessarily a symbol. If it stands for something, it becomes a symbol for this something. If it does not stand for some-thing, then it becomes not a symbol but a *meaningless* sign. This applies to words just as it does to bank cheques. If one has a zero balance in the bank, but still has a cheque-book and issues a cheque, he issues a sign but not a symbol, because it does not stand for anything. The penalty for such use of these particular signs as symbols is usually jailing. This analogy applies to the oral noises we make, which occasionally become symbols and at other times do not; as yet, no penalty is exacted for such a fraud.

Before a noise., may become a symbol, something must exist for the symbol to symbolize. So the first problem of symbolism should be to investigate the problem of 'existence'. To define 'existence', we have to state the standards by which we judge existence. At present, the use of this term is not uniform and is largely a matter of convenience. Of late, mathematicians have discovered a great deal about this term. For our present purposes, we may accept two kinds of existence: (1) the physical existence, roughly connected with our 'senses' and persistence, and (2) 'logical' existence. The new researches in the foundations of mathematics, originated by Brouwer and Weyl, seem to lead to a curtailment of the meaning of 'logical' existence in quite a sound direction; but we may provisionally accept the most general meaning, as introduced by Poincaré He defines 'logical' existence as a statement free from self-contradictions. Thus, we may say that a 'thought' to be a 'thought' must not be self-contradictory. A self-contradictory statement is meaningless; we can argue either way without reaching any valid results. We say, then, that a self-contradictory statement has no 'logical' existence. As an example, let us take a statement about a square circle. This is called a contradiction in terms, a non-sense, a meaningless statement, which has no 'logical' existence. Let us label this 'word salad' by a special noise—let us say, 'blah-blah'. Will such a noise become a word, a symbol? Obviously not—it stands for nothing; it remains a mere noise., no matter if volumes should be written about it.

It is extremely important, semantically, to notice that not all the noises., we humans make should be considered as symbols or valid words. Such empty noises., can occur not only in direct 'statements', but also in 'questions'. Quite obviously, 'questions' which employ noises., instead of words, are not significant questions. They ask nothing, and cannot be answered. They are, perhaps, best treated by 'mental' pathologists as symptoms of delusions, illusions, or hallucinations. In asylums the noises., patients make are predominantly meaningless, as far as the external world is concerned, but *become symbols in the illness of the patient*.

A complicated and difficult problem is found in connection with those symbols which have meaning in one context and have no meaning in another context. Here we approach the question of the application of 'correct symbolism to facts'. We will not now enlarge upon this subject, but will only give, in a different wording, an illustration borrowed from Einstein. Let us take anything; for example, a pencil. Let us assume that this physical object has a temperature of 60 degrees. Then the 'question' may be asked: 'What is the temperature of an "electron" which goes to make up this pencil?' Different people, many scientists and mathematicians included, would say: '60 degrees'; or any other number. And, finally, some would say: 'I do not know'. All these answers have one common characteristic; namely, that they are senseless; for they try to answer a meaningless question. Even the answer, 'I do not know', does not escape this classification, as there *is nothing to know about a meaningless question*. The only correct answer is to explain that the 'question' has no meaning. This is an example of a symbol which has no application to an 'electron'. Temperature by *definition* is the vibration of molecules (atoms are considered as mon-atomic molecules), so to have temperature at all, we must have at least two molecules. Thus, when we take one molecule and split it into atoms and electrons, the term 'temperature' does not apply by definition to an electron at all. Although the term 'temperature' represents a perfectly good symbol in one context, it becomes a meaningless noise in another. The reader should not miss the plausibility of such gambling on words, for there is a very real semantic danger in it.

In the study of symbolism, it is unwise to disregard the knowledge we gather from psychiatry. The so-called 'mentally' ill have often a very obvious and well-known semantic mechanism of projection. They project their own feelings, moods, and other structural implications on the outside world, and so build up delusions, illusions, and hallucinations, believing that what is going on in them is going on *outside* of them. Usually, it is impossible to convince the patient of this error, for his whole illness is found in the semantic disturbance which leads to such projections.

In daily life we find endless examples of such semantic projections, of differing affective intensity, which projections invariably lead to consequences more or less grave. The structure of such affective projections will be dealt with extensively later on. Here we need only point out that the problems of 'existence' are serious, and that any one who claims that something 'exists' outside of his skin must show it. Otherwise, the 'existence' is found only inside of his skin—a psycho-logical state of affairs which becomes pathological the moment he projects it on the outside- world. If one should claim that the term 'unicorn' is a symbol, he must state what this symbol stands for. It might be said that 'unicorn', as a symbol, stands for a *fanciful* animal in heraldry, a statement which happens to be true. In such a sense the term 'unicorn' becomes a symbol for a fancy, and rightly belongs to psycho-logics, which deals with human fancies, but does not belong to zoology, which deals with actual animals. But if one should believe firmly and intensely that the 'unicorn' represents an actual animal which has an external existence, he would be either mistaken or ignorant, and could be convinced or enlightened; or, if not, he would be seriously ill. We see that in this case, as in many others, all depends to what 'ology' our semantic impulses assign some 'existence'. If we assign the 'unicorn' to psycho-logics, or to heraldry, such an assignment is correct, and no semantic harm is done; but if we assign a 'unicorn' to zoology; that is to say, if we believe that a 'unicorn' has an objective and not a fictitious existence, this *s.r* might be either a mistake, or ignorance, and, in such a case, it could be corrected; otherwise, it becomes a semantic illness. If, in spite of all contrary evidence, or of the lack of all positive evidence, we hold persistently to the belief, then the affective components of our *s.r* are so strong that they are beyond normal control. Usually a person holding such affective beliefs is seriously ill, and, therefore, no amount of evidence can convince him.

We see, then, that it is not a matter of indifference to what 'ology' we assign terms, and some assignments may be of a pathological character, if they identify psycho-logical entities with the outside world. Life is full of such dramatic identifications, and it would be a great step forward in semantic hygiene if some 'ologies'—e.g., demonologies of different brands, should be abolished as such, and their subject-matter transferred to another 'ology'; namely, to psycho-logics, where it belongs.

The projection mechanism is one fraught with serious dangers, and it is very dangerous to develop it. The danger is greatest in childhood, when silly teachings help to develop this semantic mechanism, and so to affect, in a pathological way, the physically undeveloped nervous system of the human child. Here we meet an important fact which will become prominent later—that ignorance, identification, and pathological delusions, illusions, and hallucinations, are dangerously akin, and differentiated *only* by the 'emotional' background or intensity.

An important aspect of the problem of existence can be made clear by some examples. Let us recall that a noise or written sign, to become a symbol, must stand for *something*. Let us imagine that you, my reader, and myself are engaged in an argument. Before us, on the table, lies something which we usually call a box of matches: you argue that there are matches in this box; I say that there are no matches in it. Our argument can be settled. We open the box and look, and both become convinced. It must be noticed that in our argument we used *words*, because they stood for something; so when we began to argue, the argument could be solved to our mutual satisfaction, since there was a *third* factor, the object, which corresponded to the symbol used, and this settled the dispute. A third factor was present, and agreement became possible. Let us take another example. Let us try to settle the problem: 'Is blah-blah a case of tra-tra?' Let us assume that you say 'yes', and that I say 'no'. Can we reach any agreement? It is a real tragedy, of which life is full, that such an argument cannot be solved at all. We used noises, not words. There was *no third* factor for which these noises stood as symbols, and so we could argue endlessly without any possibility of agreement. That the noises may have stood for some *semantic disturbance* is quite a different problem, and in such a case a psycho-pathologist should be consulted, but arguments should stop. The reader will have no difficulty in gathering from daily life other examples, many of them of highly tragic character.

We see that we can reach, even here, an important conclusion; namely, that, first of all, we must distinguish between words, symbols which symbolize something, and noises, not symbols, which have no meaning (unless with a pathological meaning for the physician); and, second, that if we use words (symbols for something), all disputes can be solved sooner or later. But, in cases in which we use noises as if they were words, such disputes can *never* be settled. Arguments about the 'truth' or 'falsehood' of statements containing noises are useless, as the terms 'truth' or 'falsehood' do not apply to them. There is one characteristic about noises which is very hopeful. If we use words, symbols, not-noises, the problems may be complicated and difficult; we may have to wait for a long time for a solution; but we know that a solution will be forthcoming. In cases where we make noises, and treat them as words, and this fact is exposed, then the 'problems' are correctly recognized at once as 'no-problems', and such solutions remain valid. Thus, we see that one of the obvious origins of human disagreement lies in the use of noises for words, and so, after all, this important root of human dissension might be abolished by proper education of *s.r* within a single generation. Indeed, researches in symbolism and *s.r* hold great possibilities. We should not be surprised that we find meaningless noises in the foundation of many old 'philosophies', and that from them arise most of the old 'philosophical' fights and arguments. Bitterness and tragedies follow, because many 'problems' become 'no-problems', and the discussion leads nowhere. But, as material for psychiatrical studies, these old debates may be scientifically considered, to the great benefit of our understanding.

We have already mentioned the analogy between the noises we make when these noises do not symbolize anything which exists, and the worthless 'cheques' we give when our bank balance is zero. This analogy could be enlarged and compared with the sale of gold bricks, or any other commercial deal in which we try to make the other fellow accept some thing by a representation which is contrary to fact. But we do not realize that when we make noises which are not words, because they are not symbols, and give them to the other fellow as if they were to be considered as words or symbols, we commit a similar kind of action. In the concise *Oxford Dictionary of Current English*, there is a word, 'fraud', the definition of which it will be useful for us to consider. Its standard definition reads: 'Fraud, n. Deceitfulness (rare), criminal deception, *use of false representations*. (in Law, . . .); dishonest artifice or trick *(pious* fraud, deception intended to benefit deceived, and especially to strengthen religious belief); person or thing not fulfilling expectation or description.'* Commercialism has taken good care to prevent one kind of symbolic fraud, as in the instances of spurious cheques and selling gold bricks or passing counterfeit money. But, as yet, we have not become intelligent enough to realize that another most important and similar kind of fraud is continually going on. So, up to the present, we have done nothing about it.

No reflecting reader can deny that the passing off, on an unsuspecting listener, of noises for words, or symbols, must be classified as a fraud, or that we pass to the other fellow contagious semantic disturbances. This brief remark shows, at once, what serious ethical and social results would follow from investigation of correct symbolism.

On one side, as we have already seen, and as will become increasingly evident as we proceed, our *sanity* is connected with correct symbolism. And, naturally, with the increase of sanity, our 'moral' and 'ethical' standards would rise. It seems useless to preach metaphysical 'ethics' and 'morals' if we have no standards for sanity. A fundamentally *un*-sound person cannot, in spite of any amount of preaching, be either 'moral' or 'ethical'. It is well known that even the most good-natured person becomes grouchy or irritable when ill, and his other semantic characteristics change in a similar way. The abuse of symbolism is like the abuse of food or drink: it makes people ill, and so their reactions become deranged.

But, besides the moral and ethical gains to be obtained from the use of correct symbolism, our economic system, which is based on symbolism and which, with ignorant commercialism ruling, has mostly degenerated into an abuse of symbolism (secrecy, conspiracy, advertisements, bluff, 'live-wire agents'.,), would also gain enormously and become stable. Such an application of correct symbolism would conserve a tremendous amount of nervous energy now wasted in worries, uncertainties., which we are all the time piling upon ourselves, as if bent upon testing our endurance. We ought not to wonder that we break down individually and socially. Indeed, if we do not become more intelligent in this field, we shall inevitably break down racially.

The semantic problems of correct symbolism underlie *all human* life. Incorrect symbolism, similarly, has also tremendous semantic ramifications and is bound to undermine any possibility of our building a structurally *human* civilization. Bridges cannot be built and be expected to endure if the cubic masses of their anchorages and abutments are built according to formulae applying to *surfaces*. These formulae are structurally different, and their confusion with the formulae of volumes must lead to disasters. Similarly, we cannot apply generalizations taken from cows, dogs, and other animals to man, and expect the resultant social structures to endure.

*The first italics are mine. — A.K

Of late, the problems of meaninglessness are beginning to intrigue a number of writers, who, however, treat the subject without the realization of the multiordinal, ∞-valued, and *non-el* character of meanings. They assume that 'meaningless' has or may have a definite general content or unique, one-valued 'meaning'. What has been already said on *non-el* meanings, and the example of the unicorn given above, establish a most important semantic issue; namely, that what is 'meaningless' in a given context on one level of analysis, may become full of sinister meanings on another level when it becomes a symbol for a *semantic disturbance*. This realization, in itself, is a most fundamental semantic factor of our reactions, without which the solution of the problems of sanity becomes extremely difficult, if at all possible.

5. ON FUNCTION

The famous mathematician, Heaviside, mentions the definition of quaternions given by an American schoolgirl. She defined quaternions as 'an ancient religious ceremony'. Unfortunately, the attitude of many mathematicians justified such a definition. The present work departs widely from this religious attitude and treats mathematics simply as a most important and unique form of human behaviour. There is nothing sacred about any single verbal formulation, and even those that now seem most fundamental should be held subject to structural revision if need should arise. The few mathematicians who have produced epoch-making innovations in mathematical method had this behaviouristic attitude *unconsciously*, as will be shown later. The majority of mathematicians take mathematics as a clear-cut entity, 'by itself'. This is due, first, to a confusion of orders of abstractions and to identification, as will be explained later; and, second, to its seeming simplicity. In reality, such an attitude introduces quite unexpected complications, leading to mathematical revolutions, which are always bewildering. The mathematical revolutions occur only because of this *over-simplified*, and thus fallacious, attitude of the mathematicians toward their work. Had all mathematicians the semantic freedom of those who make the mathematical 'revolutions', there would be *no* mathematical 'revolutions', but an extremely swift and constructive progress. To re-educate the *s.r* of such mathematicians, the problem of the psycho-logics of mathematics must receive more attention. This means that some mathematicians must become psycho-logicians also, or that psycho-logicians must study mathematics.

For, let us take a formula which exemplifies mathematics at its best; namely, one and one make two $(1 + 1 = 2)$. We see clearly that this human product involves a threefold relation: between the man who made it,

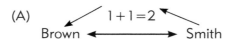

(A) $1 + 1 = 2$
Brown ⟷ Smith

let us say, Smith, and the black-on-white marks (A), between these marks and Brown, and between Brown and Smith. This last relationship is the only *important* one. The marks (A) are only auxiliary and are *meaningless by 'themselves'*. They would never occur if there were no Smiths to make them, and would be of no value if there were no understanding Browns to use and to appreciate them. It is true that when we take into account this threefold relation the analysis becomes more difficult and must involve a revision of the foundations of mathematics. Although it is impossible to attempt in this book a deeper analysis of these problems in a general way, yet this behaviouristic attitude follows the rejection of the 'is' of identity, and is applied all through this work.

The notion of 'function' has played a very great role in the development of modern science, and is structurally and semantically fundamental. This notion was apparently first introduced into mathematical literature by Descartes. Leibnitz introduced the term. The notion of a 'function' is based on that of a *variable*. In mathematics, a variable is used as an ∞-valued symbol that can represent *any one* of a series of numerical elements.

Variables are usually symbolized by the end letters of the alphabet, $x, y, z,$. The supply is increased as desired by the use of indices; for instance, $x', y', z'.; x'', y'', z''.;$ or $x_1, y_1, z_1.; x_2, y_2, z_2,$. This gives a flexible means of denoting numerous individuals, and so manufacturing them indefinitely, as the extensional method of mathematics requires. Another method, introduced not long ago, has proven useful in dealing with a definite selection of variables in a simplified manner.

One letter or one equation can be used instead of many. The variable sign x is modified by another letter which may have different values, in a given range; for instance, x_i, x_k. The modifying letter i or k can take the serial values; let us say i or $k=1, 2, 3,$. Since the one symbol x_k stands for the array of many *different* variables x_1, x_2, x_3, statements can be greatly simplified, and yet preserve structurally the *extensional* individuality.

It is important that the non-mathematical reader should become acquainted with the above methods and notations, as they involve a profound and far-reaching structural and *psycho-logical* attitude, useful to *everybody*, involving most fundamental s.r.

The *extensional* method means dealing structurally with many *definite individuals*; as, for instance, with 1, 2, 3., a series in which each individual has a special and *unique* name or symbol. This extensional method is structurally the *only* one by which we may expect to acquire [non-A] ∞-valued s.r. In a strict sense, the problems in life and the sciences do not differ structurally from this mathematical problem. In life and science, one deals with many, actual, unique individuals, and all *speaking* is using abstractions of a very high order (abstractions from abstraction from abstraction, .). So, whenever we speak, the individual is never completely covered, and some characteristics are left out.

A rough definition of a function is simple: y is said to be a function of x, if, when x is given, y is determined. Let us start with a simple mathematical illustration: $y=x+3$. If we select the value 1 for x our $y=1+3=4$. If we select $x=2$, then $y=2+3=5,$. Let us take a more complicated example; for instance: $y=x^2-x+2$. We see that for $x=1$, $y=1-1+2=2$; for $x=2$, $y=4-2+2=4$; for $x=3$, $y=9-3+2=8,$.

In general, y is determined when we fulfill all the indicated *operations* upon the variable x, and so get the final results of these operations. In symbols, $y = f(x)$, which is read, y equals function of x, or y equals f of x.

In our example, we may call x the independent variable, meaning that it is the one to which we may assign any value at our pleasure, if not limited by the conditions of our problem, and y would then be the dependent variable, which means that its value is no longer dependent on our pleasure, but is determined by the selection of the value of x. The terms dependent and independent variables are not absolute, for the dependence is mutual, and we could select either variable as the independent one, according to our wishes.

The notion of a 'function' has been generalized by Bertrand Russell to the very important notion of a 'propositional function'.[1] For my purpose, a rough definition will be sufficient. By a propositional function, I mean an ∞-*valued* statement, containing one or more variables, such that when single values are assigned to these variables the expression becomes, in principle, a *one-valued* proposition. The ∞-valued character of propositional functions seems essential, because we may have a one-valued descriptive function with variables, or a one-valued expression formulating a semantic relational law expressed in variable terms., yet these would be propositions. Thus, the ∞-*valued* statement, 'x is black', would exemplify a propositional function; but the one-valued relation 'if x is more than y, and y is more than z, then x is more than z' exemplifies a proposition. This extended *m.o* notion of a propositional function becomes of crucial importance in a [non-A]-system, because most of our speaking is conducted in [non-A] ∞-valued languages to which we mostly and delusionally ascribe single values, entirely preventing proper evaluation.

An important characteristic of a propositional function, for instance, 'x is black', is that such a statement is neither true nor false, but ambiguous. It is useless to discuss the truth or falsehood of propositional functions, since the terms true or false cannot be applied to them. But if a definite, single value is assigned to the variable *x*, then the propositional function becomes a proposition which may be true or false. For instance, if we assign to *x* the value 'coal', and say 'coal is black', the ∞-valued propositional function has become a one-valued true proposition. If we should assign to *x* the value 'milk', and say 'milk is black', this also would make a proposition, but, in this case, false. If we should assign to *x* the value 'blah-blah', and say 'blah-blah is black', such a statement may be considered as meaningless, since it contains sounds which have *no* meaning; or we *may* say, 'the statement blah-blah is *not* black but meaningless', and, therefore, the proposition 'blah-blah is black', is *not* meaningless but false.

The problems of 'meaning' and 'meaningless' are of great semantic importance in daily life, but, as yet, little has been done, and little research made, to establish or discover valid criteria. To prove a given statement false is often laborious, and sometimes impossible to do so, because of the undeveloped state of knowledge in that field. But with meaningless verbal forms, when their meaninglessness is exposed in a given case, the non-sense is exploded for good.

From this point of view, it is desirable to investigate more fully the mechanism of our symbolism, so as to be able to distinguish between statements which are false and verbal forms which have no meanings. The reader should recall what was said about the term 'unicorn', used as a symbol in heraldry and, eventually, in 'psychology', since it stands for a human *fancy*, but, in zoology, it becomes a noise and not a symbol, since it does not stand for any actual animal whatsoever.

A very curious semantic characteristic is shared in common by a propositional function and a statement containing meaningless noises; namely, that neither of them can be true or false. In the old [A] way all sounds man made, which could be written down and looked like words, were considered words; and so every 'question' was expected to have an answer. When spell-marks (noises which can be spelled) were put together in a specified way, each combination was supposed to say something, and this statement was supposed to be true or false. We see clearly that this view is not correct, that, in addition to words, we make noises (spell-marks) which may have the appearance of being words, but should *not* be considered as words, as they say nothing in a given context. Propositional functions, also, cannot be classified under the simple two opposites of true and false.

The above facts have immense semantic importance, as they are directly connected with the possibility of human agreement and adjustment. For upon statements which are neither true nor false we can always disagree, if we insist in applying criteria which have no application in such cases.

In *human* life the semantic problems of 'meaninglessness' are fundamental for sanity, because the evaluation of noises, which do not constitute symbols in a given context as symbols in that context, must, of necessity, involve delusions or other morbid manifestations.

The solution of this problem is simple. Any noises or signs, when used semantically as symbols, *always* represent *some symbolism*, but we must find out to what field the given symbolism applies. We find only three-possible fields. If we apply a symbol belonging to one field to another field, it has very often no meaning in this latter. In the following considerations, the theory of errors is disregarded.

A symbol may stand for: (1) Events outside our skin, or inside our skin in the fields belonging to physics, chemistry, physiology, . (2) Psycho-logical events inside our skin, or, in other words, for *s.r* which may be considered 'sane', covering a field belonging to psycho-logics. (3)

Semantic disturbances covering a pathological field belonging to psychiatry.

As the above divisions, together with their interconnections, cover the field of human symbolism, which, in 1933, have become, or are rapidly becoming, *experimental* sciences, it appears obvious that older 'metaphysics' of every description become illegitimate, affording only a very fertile field for study in psychiatry.

Because of *structural* and the above *symbolic* considerations based on [non-A] negative, non-identity premises, these conclusions appear as *final;* and, perhaps, for the first time bring to a focus the age-long problem of the subject-matter, character, value, and, in general, the status of the older 'metaphysics' in human economy. From the *non-el,* structural, and semantic point of view, the problems with which the older 'metaphysics' and 'philosophy' dealt, should be divided into two quite definite groups. One would include 'epistemology', or the theory of knowledge, which would ultimately merge with scientific and *non-el* psycho-logics, based on general semantics, structure, relations, multi-dimensional order, and the quantum mechanics of a given date; and the rest would represent semantic disturbances, to be studied by a generalized up-to-date psychiatry.

Obviously, considerations of structure, symbolism, sanity., involve the solutions of such weighty problems as those of 'fact', 'reality', 'true', 'false'., which are completely solved only by the consciousness of abstracting, the multiordinality of terms., —in general, a [non-A]-system.

Let me repeat the rough definition of a propositional function—as an ∞-valued statement containing variables and characterized by the fact that it is ambiguous, neither true nor false.

How about the terms we deal with in life? Are they all used as one-valued terms for constants of some sort, or do we have terms which are inherently ∞-valued or variable? How about terms like 'mankind', 'science', 'mathematics', 'man', 'education', 'ethics', 'politics', 'religion', 'sanity', 'insanity', 'iron', 'wood', 'apple', 'object', and a host of other terms? Are they labels for one-valued constants or labels for ∞-valued stages of processes. Fortunately, here we have no doubt.

We see that a large majority of the terms we use are names for ∞-valued stages of processes with a *changing content.* When such terms are used, they generally carry different or many contents. The terms represent ∞-valued variables, and so the statements represent ∞-valued propositional functions, not one-valued propositions, and, therefore, in (principle, are neither true nor false, but ambiguous.

Obviously, before such propositional functions can become propositions, and be true or false, single values must be assigned to the variables by some method. Here we must select, at least, the use of co-ordinates. In the above cases, the 'time' co-ordinate is sufficient. Obviously, 'science 1933' is quite different from 'science 1800' or 'science 300 B.C'.

The objection may be made that it would be difficult to establish means by which the use of co-ordinates could be made workable. It seems that this might involve us in complex difficulties. But, no matter how simple or how complex the means we devise, the details are immaterial and, therefore, we can accept the roughest and simplest; let us say, the year, and usually no spatial co-ordinates. The invaluable semantic effect of such an innovation is *structural,* one-, versus ∞-valued, *psycho-logical* and methodological, and affects deeply our *s.r.*

From time immemorial, some men were supposed to deal in one-valued 'eternal verities'. We called such men 'philosophers' or 'metaphysicians'. But they seldom realized that all their 'eternal verities' consisted only of *words,* and words which, for the most part, belonged to a primitive language, reflecting in its structure the assumed structure of the world of remote antiquity. Besides, they did not realize that these 'eternal verities' last only so long as the human nervous system is not altered. Under the influence of these 'philosophers', two-valued 'logic', and confusion of orders of abstractions, nearly all of us contracted a firmly rooted predilection for 'general' statements—'universals', as they were called—which, in most cases, inherently involved the semantic one-valued conviction of validity for all 'time' to come.

If we use our statements with a date, let us say 'science 1933', such statements have a profoundly modified structural and psycho-logical character, different from the old general legislative semantic mood. A statement concerning 'science 1933', whether correct or not, has no element of semantic conviction concerning 1934.

We see, further, that a statement about 'science 1933' might be quite a definite statement, and that if the person is properly informed, it probably would be true. Here we come in contact with the structure of one of those human semantic impasses which we have pointed out. We humans, through old habits, and because of the inherent structure of human knowledge, have a tendency to make static, definite, and, in a way, absolutistic one-valued statements. But when we fight absolutism, we quite often establish, instead, some other dogma equally silly and harmful. For instance, an active atheist is psycho-logically as unsound as a rabid theist.

A similar remark applies to practically all these opposites we are constantly establishing or fighting for or against. The present structure of human knowledge is such, as will be shown later, that we tend to make definite statements, static and one-valued in character, which, when we take into account the present pre-, and [A] one-, two-, three-valued affective components, inevitably become absolutistic and dogmatic and extremely harmful.

It is a genuine and fundamental semantic impasse. These static statements are very harmful, and yet they cannot be abolished, for the present. There are even weighty reasons why, without the formulation and application of ∞-valued semantics, it is not possible (1933) to abolish them. What can be done under such structural circumstances? Give up hope, or endeavour to invent methods which cover the discrepancy in a satisfactory (1933) way? The analysis of the psycho-logics of the mathematical propositional function and [non-A] semantics gives us a most satisfactory structural solution, necessitating, among others, a four-dimensional theory of propositions.

We see (1933) that we can make definite and *static* statements, and yet make them semantically *harmless*. Here we have an example of abolishing one of the old [A] tacitly-assumed 'infinities'. The old 'general' statements were supposed to be true for 'all time'; in quantitative language it would mean for 'infinite numbers of years'. When we use the date, we reject the fanciful tacit [A] 'infinity' of years of validity, and limit the validity of our statement by the date we affix to it. Any reader who becomes accustomed to the use of this structural device will see what a tremendous semantic difference it makes psycho-logically.

But the above does not exhaust the question structurally. We have seen that when we speak about ∞-valued processes, and stages of processes, we use variables in our statements, and so our statements are not propositions but propositional functions which are not true or false, but are ambiguous. But, by assigning single values to the variables, we make propositions, which might be true or false; and so investigation and agreement become possible, as we then have something definite to talk about.

A fundamental structural issue arises in this connection; namely, that in doing this (assigning single values to the variables), our attitude has automatically changed to an extensional one. By using our statements with a date, we deal with definite issues, on record, which we can study, analyse, evaluate., and so we make our statements of an extensional character, with all cards on the table, so to say, at a given date. Under such extensional and limited conditions, our statements then become, eventually, propositions, and, therefore, true or false, depending on the amount of information the maker of the statements possesses. We see that this criterion, though difficult, is feasible, and makes agreement possible.

A structural remark concerning the [A]-system may not be amiss here. In the [A]-system the 'universal' proposition (which is usually a propositional function) always implies *existence*. In [A] 'logic', when it is said that 'all A's are B', it is assumed that there are A's. It is obvious that always assuming existence leaves no place for non-existence; and this is why the old statements were supposed to be true or false. In practical life, collections of noises (spell-marks) which look like words, but which are not, are often not suspected of being meaningless, and action based on them may consequently entail unexplicable disaster. In our lives, most of our miseries do not originate in the field where the terms 'true' and 'false' apply, but in the field where they *do not apply*; namely, in the immense region of propositional functions and meaninglessness, where agreement must fail.

Besides, this sweeping and unjustified structural assumption makes the [A]-system *less general*. To the statement, 'all A's are B.' the mathematician adds 'there may or may not be A's'. This is obviously *more general*. The old pair of opposites, true and false, may be enlarged to three possibilities—statements which might be true, or false, and verbal forms which have the appearance of being statements and yet have no meaning, since the noises used were spell-marks, *not symbols* for anything with actual or 'logical' existence.

Again a [non-A]-system shares with the [non-E] and [non-N] systems a useful and important methodological and structural innovation; namely, it limits the validity of its statements, with weighty semantic beneficial consequences, as it tends from the beginning to eliminate undue, and often intense, dogmatism, categorism, and absolutism. This, on a printed page, perhaps, looks rather unimportant, but when *applied*, it leads to a fundamental and structurally beneficial alteration in our semantic *attitudes* and behaviour.

In the present work, each statement is merely the best the author can make in 1933. Each statement is given *definitely*, but with the semantic *limitation* that it is based on the information available to the author in 1933. The author has spared no labour in endeavouring to ascertain the state of knowledge as it exists in the fields from which his material is drawn. Some of this information may be incorrect, or wrongly interpreted. Such errors will come to light and be corrected as the years proceed.

It is necessary to remember that the organism works as-a-whole. In the old days we had a comforting delusion that science was a purely 'intellectual' affair. This was an *el* creed which was structurally false to facts. It would probably be below the dignity of an older mathematician to analyse the 'emotional' values of some piece of mathematical work, as, for instance, of the 'propositional function'. But such a mathematician probably never heard of psychogalvanic experiments, and how his 'emotional curve' becomes expressive when he is solving some mathematical problem.

In 1933, we are not allowed to follow the older, seemingly easier, and simpler paths. In our discussion, we have tried to analyse the problems at hand as ∞-valued manifestations of human behaviour. We were analysing the doings of Smith, Brown., and the semantic components which enter into these forms of behaviour must be especially emphasized, emphasized because they were neglected. In well-balanced persons, all psycho-logical aspects should be represented and should work harmoniously. In a theory of sanity, this semantic balance and co-ordination should be our first aim, and we should, therefore, take particular care of the neglected aspects. The *non-el* point of view makes us postulate a permanent connection and interdependence between all psycho-logical aspects. Most human difficulties, and 'mental' ills, are of *non-el* affective origin, extremely difficult to control or regulate by *el* means. Yet, we now see that purely technical scientific discoveries, because structural, have unsuspected and far-reaching beneficial *affective* semantic components. Perhaps, instead of keeping such discoveries for the few 'highbrows', who never use them fully, we could introduce them as structural, semantic and *linguistic* devices into elementary schools, with highly beneficial psycho-logical results. There is really no difficulty in explaining what has been said here about structure to children and training them in appropriate *s.r.* The effect of doing so, on sanity, would be profound and lasting.

A most important extension of the notion of 'function' and 'propositional function' has been further accomplished by Cassius J. Keyser, who, in 1913, in his discussion of the multiple interpretations of postulate systems, introduced the notion of the 'doctrinal function'. Since, the doctrinal function has been discussed at length by Keyser in his *Mathematical Philosophy* and his other writings, by Carmichael, and others. Let us recall that a propositional function is defined as an ∞-valued statement, containing one or more variables, such that when single values are assigned to these variables the expression becomes a one-valued proposition. A manifold of interrelated propositional functions, usually called postulates, with all the consequences following from them, usually called theorems, has been termed by Keyser a *doctrinal function*. A doctrinal function, thus, has no specific content, as it deals with variables, but establishes *definite relations* between these variables. In principle, we can assign many single values to the variable terms and so generate many doctrines from *one* doctrinal function. In an ∞-valued [non-A]-system which eliminates identity and is based on structure, doctrinal functions become of an extraordinary importance.*

---

*For additional discussion of important doctrinal function, see full text pp.144ff; also index.

6. ON ORDER

Undefined Terms

We can now introduce a structural *non-el* term which underlies not only all existing mathematics, but also the present work. This *bridging* term has equal importance in science and in our daily life; and applies equally to 'senses' and 'mind'. The term in question is 'order', in the sense of 'betweenness'. If we say that a, b, and c are in the order a, b, c, we mean that b is *between* a and c, and we say, further, that a, b, c, has a different order from c, b, a, or b, a, c, . The main importance of numbers in mathematics is in the fact that they have a definite *order*. In mathematics, we are much concerned with the fact that numbers represent a definite ordered series or progression, 1, 2, 3, 4 , .

In the present system, the term 'order' is accepted as *undefined*. It is clear that we cannot *define* all our terms. If we start to define all our terms, we must, by necessity, soon come to a set of terms which we cannot define any more because we will have no more terms with which to define them. We see that the structure of any language, mathematical or daily, is such that we must start implicitly or explicitly with undefined terms. This point is of grave consequence. In this work, following mathematics, I explicitly start with undefined important terms.

When we use a series of names for objects, 'Smith, Brown, Jones'., we say *nothing*. We do not produce a proposition. But if we say 'Smith kicks Brown', we have introduced the term 'kicks', which is not a name for an object, but is a term of an entirely different character. Let us call it a 'relation-word'. If we analyse this term, 'kicks', further, we will find that we can define it by considering the leg (objective) of Smith (objective), some part of the anatomy of Brown (objective), and, finally, Brown (objective). We must use a further set of terms that describes how the leg of Smith 'moves' through an 'infinity' of 'places' in an 'infinity' of 'instants' of 'time', 'continuously' until it reaches Brown.

When 'a donkey kicks a donkey' there may be a broken leg; but that is, practically, the only consequence. Not so when Smith kicks Brown. Should Brown happen to be a royal or a business man in Nicaragua or Mexico, this might be considered 'a mortal offense of a great sovereign nation to another great sovereign nation', a war might follow and many non-royals or non-business men might die. When a *symbolic* class of life enters the arena, semantic complications may arise not existing with animals.

In the relation terms, the statement, 'Smith kicks Brown', has introduced still further symbolic complications. It involves a full-fledged metaphysics, as expressed in the terms 'moves', 'infinity', 'space', 'time', 'continuity', and what not. It must be emphasized that *all* human statements, savage or not, involve a structural metaphysics.

These relational terms should be elucidated to the utmost. Until lately, the 'philosophers', in their 'Jehovah complex', usually said to the scientists: 'Hands off; those are superior problems with which only we chosen ones can deal'. As a matter of history, 'philosophers' have not produced achievements of any value in the structural line. But the 'mere' scientists, mainly mathematicians and mathematical physicists, have taken care of these problems with extremely important structural (1933) results. In the solutions of these semantic problems, the term 'order' became paramount.

Perhaps this example of an analysis of the statement, 'Smith kicks Brown', shows the justice of the contention of this work that no man can be 'intelligent' if he is not acquainted with these new works and their structural elucidations.

We see that no statement made by man, whether savage or civilized, is free from some kind of structural metaphysics involving *s.r.* We see also that when we explicitly start with *undefined* words, these undefined words have to be taken on faith. They represent some kind of implicit creed, or metaphysics, or structural assumptions. We meet here with a tremendously beneficial semantic effect of modern methods, in that we deliberately state our undefined terms. We thus divulge our creeds and metaphysics. In this way, we do not blind the reader or student. We invite criticism, elaboration, verification, evaluation., and so accelerate progress and make it easier for others to work out issues.

That we must all start with undefined terms, representing blind creeds which cannot be elucidated further *at a given moment*, may fill the hearts of some metaphysicians with joy. 'Here', they might say, 'we have the goods on the scientists; they criticize us and reject our theories, and yet they admit that they also must start with blind creeds. Now we have full justification for assuming whatever we want to.' But this joy would be short-lived for any reasonably sane individual. In mathematics, we deliberately assume the minimum, and not the maximum, as in metaphysics. The undefined terms selected for use are the *simplest* of our experience; for instance, 'order' (betweenness). Also, in experimental sciences, we assume the least possible. We demand from a *scientific* theory, according to the standards of 1933, that it should account for all relevant facts known in 1933 and should serve as a basis for the *prediction* of new facts, which can be checked by new experiments. If metaphysicians and 'philosophers' would comply with such scientific standards, their theories would be scientific. But their old theories would have to be abandoned and their new theories would become branches of science. Under such structural circumstances, there is no possibility of going *outside* of science, as we can enlarge the bounds of science without known limits, in search for structure.

It must be pointed out that no set of undefined terms is ultimate. A set remains undefined only until some genius points out simpler and more general or structurally more satisfactory undefined terms, or can reduce the number of such terms. Which set we accept is determined, in the main, by pragmatic, practical, and structural reasons. Out of two systems which have many characteristics of usefulness., in common, we would and should select the one which assumes least, is the simplest, and carries the furthest. Such changes from one set to another, when scientific, are usually epoch-making, as exemplified in mathematics.

It is important to realize that this semantic attitude signalizes a new epoch in the development of science. In scientific literature of the old days, we had a habit of demanding, 'define your terms'. The new 1933 standards of science really should be, '*state your undefined terms*'. In other words, 'lay on the table your metaphysics, your assumed structure, and then only proceed to define your terms in terms of these *undefined terms*'. This has been done completely, or approximately so, only in mathematics. Yet, probably no one will deny that the new requirements of science (1933), no matter how laborious, are really desirable, and constitute an advance in method, in accordance with the structure of human knowledge.

In the present work, this method will be employed practically all through. Of course, names for objects may be accepted without enquiry. So we have already a large vocabulary at our disposal. But names alone do not give propositions. We need *relation-words*, and it is here where our undefined terms become important. Up to this stage of the present work, I have accepted, without over-full explanation, the vocabularies made by the linguists of exact science, whom we usually call mathematicians. There is an enormous benefit in doing so, because, no matter how imperfect the mathematical vocabulary may be, it is an extensive and developed linguistic system of similar structure to the world around us and to our nervous system 1933.

Some of the most important undefined terms which play a marked role in this work are 'order' (in the sense explained), 'relation', and 'difference', although we could define relation in terms of multidimensional order. There is a remarkable structural characteristic of these terms; namely, that they are *non-el*, and that they apply to 'senses' as well as to 'mind'. It is, perhaps, well to suggest that, in future works, the terms selected should be of the *non-el* type. Since these terms apply equally to 'senses' and to 'mind', we see that in *such* terms we may attempt to give a 'coherent' account of what we experience. The expression 'coherent' implies 'mind', and 'experience' implies 'senses'. It is amusing to watch this peculiar circularity of human knowledge, many instances of which will appear later on. Thus, there was great difficulty in expressing organism-as-a-whole notions; we had to grope about in establishing the beginnings of a suitable vocabulary before we could approach problems which were antecedent in order.

It is necessary to notice a rather curious structural similarity between the [non-N] and [non-A] systems. In both cases, we deal with certain velocities about which we know positively that they are *finite*. The velocity of light in the [N]-system was assumed to be infinite, although we know it is not so, and so 'simultaneity' had absolute meaning. The [non-N] systems introduced the finite velocity of light by ordering events, which happens to be true to facts, and thus 'simultaneity' lost its absolute character. Likewise in the [A]-system and language, the velocity of nervous impulses was assumed to be infinite, to spread 'instantaneously'. And so we had most perplexing 'philosophical' rigmaroles about 'emotions', 'intellect'., taken as independent separate entities. When we introduce explicitly the finite velocity of nervous impulses (on the average, 120 metres per second in the human nerves), we are able to reach a perfectly clear understanding, *in terms of order*, of the spread of impulses. Some 'infinite velocity' does not involve *order*. Conversely, by considering the order of events, we introduce finite velocities. We shall see later that 'infinite velocity' is *meaningless* and so all actual happenings can be ordered. The above is an important factor in our *s.r.*

Let us give a rough example. Assume that Smith has had a bad dinner. Some nervous impulse, originating from the bad dinner, starts going. At this stage, we may call it an 'undifferentiated' nervous impulse. It travels with *finite* velocity, reaches the brain-stem and the approximately central part of his brain, which we call the thalamus; is affected by them and is no longer 'undifferentiated' but becomes, let us say, 'affective'. In the cortex, it is affected again by the lessons of past experiences. It returns again to the lower centres and becomes, let us say, 'emotion'; and then anything might happen, from sudden death to a glorious poem.

The reader must be warned that this example is rough and oversimplified. Impulses are reinforced and 'inhibited' from a complex chain of nervous interconnections. But what I wish to show by this example, is that, by accepting the *finite* velocity of nerve currents, in terms of *order*, we can build up a definite vocabulary to deal, not only with the 'organism-as-a-whole', but also with the different stages of the process. This is important because, without some such *ordinal* scheme, it is structurally impossible to evade enormous verbal and semantic difficulties which lead to great confusion.

Structure, Relations, and Multi-dimensional Order

In such an ordered cyclic chain, the nerve impulses reach and traverse the different levels with *finite* velocity and so, in each case, in a *definite order*. 'Intelligence' becomes a manifestation of life of the organism-as-a-whole, structurally impossible in some fictitious 'isolation'. To 'be' means *to be related.* To be related involves multi-dimensional *order* and results in *structure*.

'Survival', 'adaptation', 'response', 'habit formation', 'orientation', 'learning', 'selection', 'evaluation', 'intelligence', 'semantic reactions', and all similar terms involve structurally an ordered, interrelated structural complex in which and by which we live and function. To 'comprehend', to 'understand', to 'know', to be 'intelligent'., in the pre-human as well as the human way, means the most useful survival adjustment to such an ordered, interrelated structure as the world and ourselves.

In this vocabulary, 'structure' is the highest abstraction, as it involves a whole, taken as-a-whole, made up of *interrelated parts*, the relations of which can be defined in still simpler terms of order. 'Knowing', in its broadest as well as in its narrow human sense, is conditioned by structure, and so consists of *structural* knowings. All empirical structures involve relations, and the last depend on multi-dimensional order. A language of *order*, therefore, is the simplest form of language, yet in structure it is similar to the structure of the world and ourselves. Such a language is bound to be useful for adaptation and, therefore, sanity; it results in the understanding of the structural, relational, multi-dimensional order in the environment on all levels.

We must stress the structural fact that the introduction of *order* as a fundamental term abolishes some fanciful and semantically very harmful 'infinities'. If an impulse could travel in 'no time' or with 'infinite velocity', which is a *structural impossibility* in this world, such an impulse would reach different places 'instantaneously', and so there would be *no order involved*. But, as soon as the actual order in which impulses reach their destination is found, 'infinite velocities' are abolished. We shall show later that 'infinite velocity' is a meaningless noise; here we stress only the point that it is a structural impossibility, as structure involves relations and orders, and order could not exist in a world where 'infinite velocities' were possible.

Conversely, if in our analysis we disregard order, we are bound to disregard relations and structure and to introduce, by necessity, some fanciful 'infinite velocities'. Any one who treats 'mind' in 'isolation' makes a structurally false assumption, and, by necessity, unconsciously ascribes some meaningless 'infinite velocity' to the nerve currents.

We have dwelt upon this subject at such length because of its general structural and semantic importance. The first step towards understanding the theory of Einstein is to be entirely convinced on the above points. Newton's disregard of order introduced an unconscious false to facts assumption of the 'infinite velocity' of light, which fatalistically leads to an objectification called 'absolute time', 'absolute simultaneity', and so introduces a terminology of inappropriate structure. A similar remark applies to arguments about 'mind' in an objectified, 'isolated' way. These arguments disregard the *order* in which the nervous impulses spread and so, by necessity, introduce a silent false to facts assumption of the 'infinite velocity' of nerve currents.

On empirical structural grounds, we know neurological and general facts on two levels. (1) Macroscopically, we have a structure in levels, stratified, so to say, with complexities arising from the general colloidal physico-chemical structure of the organism-as-a-whole. (2) The general sub-microscopic, atomic, and sub-atomic structure of all materials simply gives us the persistence of the macroscopic characteristics as the relative *invariance of function*, due to dynamic equilibrium, and ultimately reflected and conditioned by this *sub-microscopic structure* of all materials. Under such actual structural conditions, terms like 'substance', 'material'., and 'function', 'energy', 'action'., become interconnected— largely a problem of preference or necessity of selecting the level with which we want to deal.

On sub-microscopic levels, 'iron', or anything else, means only a persistence for a limited 'time' of certain gross characteristics, representing a process (structurally a four-dimensional notion involving 'time'), which becomes a question of structure. With the 1933 known unit of the world called an 'electron', which appears as an *'energy'* factor, the relative persistence or invariance of dynamic sub-microscopic structure gives us, on macroscopic levels, an average, or statistical, persistence of gross macroscopic characteristics, which we label 'iron'.

The above should be thoroughly understood and digested. As a rule, we all identify orders and levels of abstractions and so have difficulty in keeping them separate verbally (and, therefore, 'conceptually'). Thorough structural understanding helps us greatly to acquire these new and beneficial *s.r.*

Under such structural *empirical* conditions, a language of order, which implies relations and structure, as enlarged to the order of abstractions or level of consideration, largely volitional, becomes the only language which, in *structure*, is similar to the structure of the world, ourselves included, and so, of necessity, will afford the maximum of semantic benefits.

It should be understood that, on structural grounds, terms like 'substance' and 'function' become, in 1933, perfectly *interchangeable*, depending on the order of abstraction. 'Substance', for instance, on the macroscopic level becomes 'invariance of function' on the sub-microscopic level. It follows that what we know about the macroscopic ('anatomical') structure can be quite legitimately enlarged by what we know of *function* (structure on different levels). This interchangeability and complementary value of evidence is conditioned by structural considerations, and the fact that 'structure' is multiordinal. On gross anatomical grounds, we know a great deal about this structure of the nervous system. Because of experimental difficulties, very little is known of the structural submicroscopic happenings, yet we can, speak about them with benefit in *functional* terms; as, for instance, of 'activation', 'facilitation', 'resistance', 'psychogenetic effects', 'diffusion', 'permeability', the older 'inhibition', .

In such a cyclic chain as the nervous system, there is, as far as energy is concerned, no last stage of the process. If there is no motor reaction or other reflex, then there is a semantic or associative reaction with 'inhibitory' or activating consequences, which are functionally equivalent to a motor reaction. At each stage, a 'terminal' receptor is a *reacting* organ in the chain.

We know quite well from psychiatry how nervous energy may deviate from constructive and useful channels into destructive and harmful channels. The energy is not lost, but misdirected or misapplied. For instance, an 'emotional' shock may make some people release their energy into useful channels, such as concentrated efforts in some direction, which would have been impossible without this shock; but, in others, an 'emotional' shock leads to the building up of morbid 'mental' or physical symptoms.

Since the nervous structure is cyclic in most of its parts, as well as as-a-whole, and since these cycles are directly or indirectly interconnected, mutual interaction of those cycles may produce most elaborate behaviour patterns, which may be spoken of, in their manifestations, in terms of *order*. As each more important nerve centre has incoming and outgoing nerve-fibres, the activation, or reinforcement, or diffusion of nerve currents may sometimes manifest itself in our *s.r* as *reversal of order* in some aspects. Neurologically, considered on the sub-microscopic level, it would only be a case of activation, or of diffusion, or of 'inhibition'., probably *never* a problem of reversed order in the actual nerve currents.

The semantic manifestations of order and reversed order are of *crucial importance*, for we are able to *train* the individual to different orders or reversals of orders. This procedure neurologically involves activations, enforcements, diffusions, 'inhibitions', resistances, and all the other types of nervous activities, which without the formulation of psychophysiology were all *most inaccessible to* direct training. The structural fact that order and reversal of order in semantic manifestation, which all are on the un-speakable objective level, have such intimate and profound connection with fundamental nervous processes, such as activation, enforcement, diffusion, permeability, 'inhibition', resistance., gives us tremendous new powers of an educational character in building up *sanity*, and supplies methods and means to affect, direct, and train nervous activities and *s.r*, which we were *not able formerly to train* psychophysiologically. Perhaps one of the main values of the present work is the discovery of physiological means, to be given presently, for training the human nervous system in 'sanity'.

The reader should be aware, when we speak of order and reversal of order, that we mean the order and reversal of order in the un-speakable *s.r*; but the neurological mechanism is of a different character, as already explained. Our analysis of the simple semantic manifestations involve *evaluation* and so *order*, permitting a most complex re-education and re-training of the nervous system, which were entirely beyond our reach with the older methods.

Experimental evidence seems to corroborate what has been suggested here, and the analysis, in terms of order, seems to have serious practical neurological significance, owing to *similarity of structure*, resulting in *evaluation*, and so appropriate *s.r.*

For our analysis in terms of order, we start with the simplest imaginable nervous cycle; but it must be explicitly understood that such simple cycles actually never exist, and that our diagrams have value only as picturing the cyclic *order*, without complications. Let us repeat that the introduction of an analysis in terms of *order* or *reversal of order* in *the manifestation* involves, under educational influence, various other and *different actual nervous activities*, a class of activities which hitherto have always evaded our educational influences. For structural purpose, it is sufficient to make use of the distinction between lower and higher centres (a rough and ready distinction) and to consider the lower centres generally in connection with the thalamus and brain-stem (perhaps also other sub-cortical layers), and the higher centres generally in connection with the cortex. This lack of precision is intentional, for we need only sufficient structural stratification to illustrate *order*, and it seems advisable to assume only the well-established *minimum* of structure.

We have already mentioned the absolute individuality of the organism and, as a matter of fact, of everything else on objective levels. The reader need have no metaphysical shivers about such extreme individuality on the un-speakable levels. In our human economy, we need both similarities and differences, but we have as yet, in our [A]-system, chiefly concentrated our attention and training on similarities, disregarding differences. In this work, we start structurally *closer to nature* with un-speakable levels, and make *differences* fundamental, similarities appearing only at a later stage (order) *as a result of higher abstractions*. In simple words, we obtain similarities by disregarding differences, by a process of abstracting. In a world of only absolute differences, without similarities, recognition, and, therefore, 'intelligence', would be impossible.

It is possible to demonstrate how 'intelligence' and abstracting both started together and are due to the physico-chemical structure of protoplasm. All living material, usually called protoplasm, has, in some degree, the nervous functions of irritability, conductivity, integration,. It is obvious that a stimulus S does not affect the little piece of primitive protoplasm A 'all over and at once' (infinite velocity), but that it affects it first in a definite spot B, that the wave of excitation spreads, with finite velocity and usually in a diminishing gradient, to the more remote portions of A. We notice also that the *effect* of the stimulus S on A is not *identical* with the stimulus itself. A falling stone is *not* identical with the pain we feel when the stone falls on our foot. Neither do our feelings furnish a full report as to the characteristics of the stone, its internal structure, chemistry, . So we see that the bit of protoplasm is affected only *partially*, and in a *specific* way, by the stimulus. Under physico-chemical conditions, as they exist in life, there is no place for any 'allness'. In life, we deal structurally only with 'non-allness'; and so the term, '*abstracting* in different orders', seems to be structurally and uniquely appropriate for describing the effects of external stimuli on living protoplasm. 'Intelligence' of any kind is connected with the abstracting (non-allness) which is characteristic of all protoplasmic response. Similarities are perceived only as differences become blurred, and, therefore, the process is one of abstracting.

The important novelty in my treatment is in the structural fact that I treat the term 'abstracting' in the *non-el* way. We find that all living protoplasm 'abstracts'. So I make the term abstracting fundamental, and I give it a wide range of meanings to correspond to the facts of life by introducing abstractions of *different orders*. Such a treatment has great structural advantages, which will be explained in Part VII.

As our main interest is in 'Smith$_n$', we will speak mostly of him, although the language we use is structurally appropriate for characterizing all life. 'Abstracting' becomes now a physiological term with structural, actional, physico-chemical, and *non-el* implications.

Accidentally, some light is thus thrown on the problem of 'evolution'. In *actual* objective life, each new cell is different from its parent cell, and each offspring is *different* from its parents. Similarities appear only as a result of the action of our nervous system, which does not register absolute differences. Therefore, we register similarities, which evaporate when our means of investigation become more subtle. Similarities are read *into nature* by our nervous system, and so are structurally less fundamental than differences. Less fundamental, but no less important, as life and 'intelligence' would be totally impossible without *abstracting*. It becomes clear that the problem which has so excited the *s.r* of the people of the United States of America and added so much to the merriment of mankind, 'Is evolution a "fact", or a "theory"?', is simply silly. Father and son are never identical—that surely is a structural 'fact'—so there is no need to worry about still higher abstractions, like 'man' and 'monkey'. That the fanatical and ignorant attack on the theory of evolution should have occurred may be pathetic, but need concern us little, as such ignorant attacks are always liable to occur. But that biologists should offer 'defences', based on the confusion of orders of abstractions, and that 'philosophers' should have failed to see this simple dependence is rather sad. The problems of 'evolution' are verbal and have nothing to do with life as such, which is made up all through of *different* individuals, 'similarity' being structurally a manufactured article, produced by the nervous system of the observer.

In my own practice, I have become painfully aware of a similar discrepancy in the learned *s.r* of some older professors of biology, who quite often try to inform me that 'Life is overlapping', and that 'no sharp distinction between "man" and "animal" can be made'. They forget or do not know that, structurally, actual 'life' is composed of *absolute individuals*, each one *unique* and different from all other individuals. Each individual should have its individual name, based on mathematical extensional methods; for instance, $x_1, x_2, x_3, \ldots x_n$; or Smith$_1$, Smith$_2$, or Fido$_1$, Fido$_2$, . 'Man' and 'animal' are only labels for verbal fictions and are not labels for an actual living individual. It is obvious that as these verbal fictions, 'man' and 'animal', are not the living individual, their 'overlapping' or 'not overlapping' depends only on *our ingenuity*, our power of observation and abstraction, and our capacity of coining *non-el* functional definitions.

Let us see how adaption might work in practice. Let us consider two or three caterpillars, which we may name C_1, C_2, C_3, since each of them is an absolute individual and different from the others. Let us assume that C_1 is positively heliotropic, which means that he is compelled to go toward light; that C_2 is negatively heliotropic, which means that he would tend to go away from light; and that C_3 is non-heliotropic, which means that light-would have no effect on him of a directional character. At a certain age, C_1 would crawl up the tree near which he was born and so reach the leaves, eat them, and, after eating them, would be able to complete his development. C_2, and probably C_3, would die, as they would not crawl up the tree toward sufficient food. Thus, we see that among the indefinite number of possible individual make-ups of C_k ($k=1, 2, 3, \ldots$ n), each one being different, only those which were positively heliotropic would survive *under the conditions of this earth*, and all the rest would die. The positively heliotropic would propagate and their positive heliotropism might be perpetuated, the negatively heliotropic and non-heliotropic becoming extinct. This would only occur, however, in a world in which trees have roots in the ground and leaves on their parts toward the sun. In a world where the trees grew with their roots toward the sun and leaves in the ground, the reverse would happen; namely, the negatively heliotropic and non-heliotropic would survive, and the positively heliotropic would die out. We can not foretell whether, in such a world, there would be caterpillars; so this is an hypothetical example.

Experiments made with such caterpillars have shown that the positively heliotropic ones crawl toward the sun, even upon a plant which has been turned over, with the roots toward the sun. They crawl away *from food*, and die. We see that the external *environmental* conditions determine which characteristics survive, and so we reach the notion of adjustment.

The practical result of these conditions is that the indefinite number of individual variations, although they undoubtedly exist, seldom come to our attention, as those variations which do not fit their environmental conditions become extinct; their variations do not become hereditary, and consequently we can seldom find them outside of a laboratory.

This shows, also, the permanent connection and interdependence of the facts of nature. The structural fact that our trees grow with their roots in the ground and their leaves upward is not an independent fact; it has something to do with the structure of the world and the position and the effect of the sun, . So the fact that we have positively heliotropic caterpillars of a special kind, and not negatively heliotropic ones, has something to do with the structure of the rest of the world.

To illustrate this interconnection and interdependence of nature still more clearly, let me suggest an hypothetical question. How would conditions, as they are on this earth, compare with those which would obtain, if it were, let us say, one mile greater in diameter? Some try to guess the answer; yet this question cannot be answered at all. The diameter of this earth is strictly dependent on all the structural conditions which prevail in this world. Since it is impossible to know what kind of a universe it would be, in which this earth could be different from what we know, it is, of course, equally impossible to foresee whether on such a fictitious earth, in such a fictitious universe, there would even be life at all. Because the structure of the world is such as we know it, our sun, our earth, our trees, our caterpillars, and, finally, ourselves have their structure and characteristics. We do not need to enter here into the problems of determinism versus indeterminism, as these problems are purely verbal, depending on our orders of abstractions, the 'logic' we accept, and so, finally, on our pleasure, as is explained more in detail later on, and could not be solved satisfactorily in an [A], *el* system, with its two-valued 'logic'.

According to the evidence at hand in 1933, 'Smith', appears among the latest inhabitants of this earth, and subject to the general test of survival, as already explained. The few thousand years during which there had been any 'Smith$_n$' are too short a period to test, with any certitude, his capacity for survival. We know of many species of animals, and also races of man, of which very little trace has been left. What we know about their history is mostly through a few fossils, which are kept in museums.

The external world is full of devastating energies and of stimuli too strong for some organisms to endure. We know that only those organisms have survived which could successfully either protect themselves from over-stimulations or else were under protective circumstances. If we look over the series of surviving individuals, paying special attention to the higher animals and man, we find that the nervous system has, besides the task of conducting excitation., the task of so-called 'inhibition'. Response to stimuli, by survival, proved its usefulness. But to diminish the response to some stimuli or avoid stimuli, proved also to be useful, again by survival. It is known that the upper or latest layers of the nervous system are mostly such *protective* layers, to prevent immediate responses to stimulation. With the development of the nervous system from the simplest to the most complex, we see an increase in behaviour of a modifiable or individually adjustable type. In terms of *order*, and using the old language, 'senses' came first in order and 'mind' next, in all their forms and degrees.

If we speak in neurological terms, we may say that the present nervous structure is such that the entering nerve currents have a natural direction, established by survival; namely, they traverse the brain-stem and the thalamus first, the sub-cortical layers next, then the cerebral cortex, and return, transformed, by various paths. Experience and science in 1933 are showing that this is the order established under a heavy toll of destruction and non-survival in a system of adjustment, and so should be considered the 'natural' order, because of its survival value.

We all know in practice about a 'sensation', and a 'mental picture' or 'idea'. As 'sensations' were often very deceptive and, therefore, did not always lead to survival, a nervous system which somehow retained vestiges, or 'memories', of former 'sensations' and could recombine them, shift them., proved of higher survival value, and so 'intelligence' evolved, from the lowest to the highest degrees.

Experience and experiments show that the natural order was 'sensation' first, 'idea' next; the 'sensation' being an abstraction of some order, and the 'idea' already an abstraction from an abstraction or an abstraction of higher order.

Experience shows again that among *humans*, this order in manifestations is sometimes *reversed;* namely, that some individuals have 'idea' first; namely, some vestiges of memories, and 'sensations' next, without any external reason for the 'sensations'. Such individuals are considered 'mentally' ill; in legal terms, they are called 'insane'. They 'see', where there is nothing to see; they 'hear', where there is nothing to hear; they are paralysed, where there is no reason to be paralyzed; they have pains, when there is no reason to have pains, and so on, endlessly. Their survival value, if not taken care of, is usually nil. This reversal of order, but in a mild degree, is extremely common at present among all of us and underlies mainly all human misfortunes and un-sanity.

This reversal of order in its mild form is involved in identification or the confusion of orders of abstractions; namely, when we act as if an 'idea' were an 'experience' of our 'senses', thus duplicating in a mild form the mechanism of the 'mentally' ill. This implies nervous disturbances, since we *violate the natural (survival) order of the activities of the nervous system*. The mechanism of *projection* is also connected with this *reversal of order*. This reversal transforms the external world into a quite different and fictitious entity. Both ignorance and the old metaphysics tend to produce these undesirable nervous effects of *reversed order and so non-survival evaluation*. If we use the nervous system in a way which is against its survival structure, we must expect non-survival. Human history is short, but already we have astonishing records of extinction.

That such reversal of order in the manifestations of the functioning of the nervous system must be extremely harmful, becomes evident when we consider that in such a case the upper layers of our nervous system (the cortex) not only do not protect us from over-stimulation originating in the external world and inside us, but actually contribute to the over-stimulation by producing fanciful, yet very real, irritants. Experiments on some patients have shown how they benefit *physically* when their internal energy is liberated from fighting phantoms and so can be redirected to fight the colloidal disturbances. Such examples could be cited endlessly from practically every field of medicine and life. This problem of 'reversal of order' is not only very important semantically, but also very complex, and it will be analysed further on.

The reader should not miss the fact that an analysis in terms of *order* throws a new light on old problems, and so the scientific benefit of the use of such a term is shown. But this is not all; the use of the term *order* has brought us to the point where we can see far-reaching practical applications of the knowledge we already possess, and of which we have not so far made any systematic use.

We know that the activity of the nervous system is facilitated by repetition, and we can learn useful habits as easily as harmful. In the special case of *s.r* also we can train ourselves either way, though one may have useful survival value; the other being harmful, with no survival value. The problem is again one of *order*, and, among others, a problem of extension and intension, as has already been mentioned several times.

Order and the Problems of Extension and Intension

The problems of extension and intension are not new, but have been treated as yet only casually by 'philosophers', 'logicians', and mathematicians, and it has not been suspected what profound, far-reaching, and important structural psycho-logical, semantic components they represent.

We usually forget that whenever a mathematician or a 'philosopher' produces a work, this involves his 'attitude', which represents an extremely complex psycho-logical *s.r* of the organism-as-a-whole. In most cases, these attitudes determine not only the character of our work, but also other reactions which make up our individual and social life. Historically, the mathematicians have a steady record of achievement, and 'philosophers' (excluding epistemologists) one of uselessness or failure. Has this record something to do with the extensional and intensional *attitudes*? In fact, it has. It is easy to show that the extensional attitude is the *only one* which is in accordance with the *survival order and nervous structure*, and that the intensional attitude is the reversal of the natural order, and, therefore, must involve non-survival or pathological *s.r*.

One of the simplest ways of approaching the problems of 'extension' and 'intension' is perhaps to point its connection with definitions. A collection may be defined, so we are told, by enumeration its members, as, for instance, when we say that the collection contains Smith, Brown, Jones, . Or we may define our collection by giving a defining 'property'. We are told that the first type of definitions which enumerates individual members is to be called a definition by *extension*, the second, which gives a defining 'property', is to be called a definition by *intension*.

We can easily see that a 'definition by extension' uniquely characterizes the collection, Smith₁, Brown₁, Jones₁,. Any other collection, Smith₂, Brown₂, Jones₂, would obviously be different from the first one, since the individuals differ. If we 'define' our collection by intension; that is, by ascribing some characteristic to each of the individuals, for instance, that they have no tails, many collections of individuals without tails might be selected. Since these collections would be composed of entirely different individuals, they would be entirely different, yet by 'intension', or defining characteristic, they all would be supposed to be one collection.

As explained before, the structure of our nervous system was established with 'senses' first, and 'mind' next. In neurological terms, the nervous impulses should be received first in the lower centres and pass on through the sub-cortical layers to the cortex, be influenced there and be transformed in the cortex by the effect of past experiences. In this transformed state they should then proceed to different destinations, as predetermined by the structure established by survival values. We know, and let us remember this, that the reversed order in semantic manifestation—namely, the *projection into 'senses' of memory traces* or doctrinal impulses—is against the survival structure, and hallucinations, delusions, illusions, and confusion of orders of abstractions are to be considered pathological. In a 'normal' human nervous system with *survival* value, the nervous impulses should not be lost in the sub-cortical layers. In such a case, the activity of our human nervous system would correspond to the activity of the less-developed nervous systems of animals which have no cortex at all. It must be remembered, also, that the sub-cortical layers which have a cortex, as in man, are quite different from corresponding layers of those animals which have never developed a cortex. It is impossible to avoid the conclusion that survival values are *sharply* characterized by *adequacy*, and that animals without cortex have nervous systems adequate for their needs under their special conditions; otherwise, they would not have survived. This applies, also, to those animals who have a cortex. Their activities for survival depend on this cortex; and when the cortex is removed, their activities become inadequate. Their sub-cortical layers alone are not adequate to insure survival. For survival, such animals must use not only their lower centres and their sub-cortical layers, but also their cortices.

Among animals, as all evidence shows, the enormous majority have, without human interference, nervous systems working usually in the 'normal' way; that is, according to the survival structure. 'Insanity' and kindred nervous disturbances are known only among ourselves (however, see Part VI). Apparently, the cortex, through its enormous internal complexity, which provides many more pathways, and through its complex interconnections, which offer many more possibilities, with a greater number of degrees of 'inhibition', of excitability, of delayed action, of activation., introduces not only a much greater flexibility of reaction, but, through this flexibility, a possibility for abuse, for reversal of manifestations, and so for a deterioration of the survival activity of the nervous system as-a-whole. The sub-cortical layers and other parts of the brain of man are different from the corresponding parts of the animal brain, which has a less-developed cortex. The nervous system works as-a-whole, and the anatomical homology of the parts of different nervous systems is a very inadequate, perhaps even a misguiding, foundation for inferring *a priori* the *functioning* of these systems, which ultimately depend not only on the macroscopic but also on the microscopic and sub-microscopic structures. For instance, we can cut off the head of some insects, and they go along quite happily and do not seem to mind the operation much. But we could not repeat this with higher animals. The behaviour is changed. A decorticated pigeon behaves differently from a decorticated rat, though neither of them seems affected greatly by the operation. A decorticated dog or ape is affected much more. Man is entirely changed. None of the higher types is able to survive long if decorticated.

There is on record the medical history, reported by Edinger and Fischer, of a boy who was born entirely without cerebral cortex. There were apparently no other important defects. This child never showed any development of sensory or motor power, or of 'intelligence', or signs of hunger or thirst. During the first year of his life, he was continually in a state of profound stupor, without any movements and from the second year on, until his death (at three years and nine months), was continually crying.

Although many animals, for instance fishes, have no differentiated cortex, yet their nervous system is perfectly *adequate* for their lives and conditions. But in a more complex nervous system, the relative functions of different parts of the brain undergo fundamental transformation. In the most complex nervous system, as found in man, the older parts of the brain are much more under the control of the cerebral cortex than in any of the animals, as is shown in the example above. The absence of the human cortex involves a much more profound disturbance of the activities of the other parts of the brain. Since the cortex has a profound influence upon the other parts of the brain, the insufficient use of the cortex must reflect detrimentally upon the functioning of the other parts of the brain. The enormous complexity of the structure of the human brain and the corresponding complexity of its functioning accounts not only for all human achievements, but also for all human difficulties. It also explains why, in spite of the fact that our anatomy differs but little from that of some higher animals, veterinary science is more simple than human medicine.

Because of the structure of the nervous system, we see how the completion of one stage of the process which originated by an external stimulus (A) and has itself become a nervously elaborated *end-product* (B), may, in its turn, become the stimulus for a still further nervously elaborated *new end-product* (C), and so on. When association or relation neurons enter, the number of possibilities is enormously increased.

It must be emphasized that A, B, C., are, fundamentally, entirely *different*. For instance, the external event A may be a falling stone, which is an entirely different affair from the pain we have when this stone falls on our foot. It thus becomes clearer what is meant by a statement that the 'senses' abstract in their own appropriate way, determined by survival value, the external events; give these abstractions their special colouring (a blow on the eye gives us the feeling of *light*); discharge these transformed stimuli to further centres, in which they become again abstracted, coloured, transformed, . The end-product of this second abstracting is again an entirely different affair from the first abstraction.

Obviously, for survival value, this extremely complex nervous system should work in complete co-ordination. Processes should pass the *entire cycle*. If not, there must be something wrong with the system. The activities of the organism are then regressive, of lower order, a condition known as 'mental' illness. The gross anatomical divisions of the nervous system should not be relied upon too much as an index of function. Perhaps these anatomical speculations are even harmful for understanding, because they falsify the facts, emphasize the macroscopic similarities unduly and disregard subtle yet fundamentally important microscopic and sub-microscopic structures and differences, which are perfectly manifest in the functioning, but which are difficult to observe directly on their level.

The term 'abstracting' is a multiordinal term, and hence has different meanings, depending on the order of abstractions. It is a functional term and, to indicate the differences in meanings, it is necessary to indicate the different orders. It is structurally a *non-el* term, built upon the extensional mathematical pattern x', x'', x'''., or x_1, x_2, x_3, . . . x_n, or x_k ($k=1, 2, 3, . . . n$). This allows us to give the term 'abstracting of different orders' a perfectly *unique* meaning in a given problem and yet to keep in a fluid state its most important *functional implications*. Something similar happens when the mathematician discusses his x_n. No one can miss the fact that he deals with a variable which can take n values; so this symbol has a quite definite descriptive structural and semantic value. So has the 'abstracting of different orders'.

It is desirable to introduce consciously and deliberately *terms* of a *structure* similar to the *structure* of human knowledge, of our nervous system, and of the world, involving appropriate *s.r.* Multiordinal terms are uniquely appropriate, since they take their ∞-valued structure from the structure of events (1933) and do not reflect their older, one-valued, false to fact character on the events. (Note the order.)

Now we are ready to reformulate the problem of extension and intension in terms of *order*.

If the natural survival order is lower abstractions first and higher next, then extension starts with absolute individuals, and conforms to the proper survival order. Extension recognizes the uniqueness, with corresponding one-value, of the individual by giving each individual a unique name, and so makes confusion impossible. Training in *s.r* of sanity becomes a possibility, and order becomes paramount. Extension and order cannot be divided. When we speak about 'order', we imply extension, and, when we speak about extension, order is implied. That modern mathematics and mathematical 'logic' has so much to do with order, as to make this term fundamental, is a necessary consequence of the extensional method which starts with unique individuals, labels them by unique names and only then generalizes or passes to ∞-valued higher order abstractions like 'numbers', . The direction of the process of abstraction is here in the survival order, from lower abstractions to higher. It hardly needs to be emphasized that, to the best of our knowledge in 1933, it is the only possible way to follow the natural order and to evade reading into a fundamentally one-valued external event, our older undi*fferentiated* ∞-valued fancies (which happens if the process is reversed in order) involving powerful factors in our *s.r.*

Intension means structurally the reversal of the survival order, since it starts with undifferentiated ∞-valued higher abstractions and distorts or disregards the essential one-values of the individuals and reads into them as *uniquely* important undifferentiated ∞-valued characteristics.

Historically, mathematicians have had a predilection and, because of the character of their 'element' (numbers) and their technique, a structural necessity, for the use of extensional methods. It does not need much imagination to see why they have produced results of utmost (although relative) importance and validity at each date.

'Philosophers' and reasoners of that class have had a predilection for intension, and this also explains why, in spite of tremendously acute verbal exercises, they have not produced anything of lasting value, for they were carried away by the structure of the language they used. This predilection being already based on the reversal of the survival order, it was bound to lead in the less-resistant individuals to nervous and 'mental' defects.

The issues, as presented here, are very clear-cut, and, in fact, too clear-cut, as we have disregarded for the present the *cyclic* order of the nervous process. This last fact first abolishes the sharp distinction between 'pure' extension and 'pure' intension, each process never being 'pure', but always 'impure', one influencing the other. 'Pure' intension and 'pure' extension are delusional, to be found only in 'mentally' ill, with no survival value. This explains why we have to use the terms of *preference* and *order*. Without these terms I would not have been able to carry through this analysis at all. This reversal of order in *s.r* implies. different distribution, diffusion, intensity of nerve currents in the submicroscopic field, and so involves important, different semantic components of non-survival value. It is most desirable to learn to control the activities on the sub-microscopic level by means of training on the macroscopic level, if means to do this can be devised.

The writer is not at all convinced that, acting as we do under the spell of intensional and ignorant 'philosophers', the existing systems and educational methods are not largely following the reversal of the survival order of our nervous processes. It seems unnecessary to point out that a structural and semantic enquiry on this particular question might be important and beneficial. It seems, without much doubt, that human institutions and activities should be in accord with 'human nature', if we are to expect them to survive without crushing us, and a scientific enquiry in this 'human nature' would be not only desirable but exceedingly important.

The reader, with the help of another person, should perform a very simple experiment. Let the assistant select secretly a dozen newspaper headlines of letters of equal size. Let the reader then sit in a chair without altering his distance from the assistant and let the assistant show him one of these headlines. If he is able to read the headline, it should be rejected, and a new one selected by the assistant and put a foot or more farther away. If this one is read correctly, it should be rejected, and a third one placed still farther away. By such trials we can finally find a distance which is slightly greater than the maximum range of clear visibility for the reader, so that, although the headline is only slightly beyond the distance at which one could read it, yet it would be illegible. Let the reader then try as hard as possible to read headlines which are just beyond his visual range. When he is convinced that he cannot read the headline, let the assistant *tell* him the content of it. Then the sitter can usually *see* with his *eyes* the letters, when he *knows* what is supposed to be there. The question arises, what part in the 'seeing' is due to 'senses', and what to 'mind'? The answer is, that, structurally, the 'seeing' is the result of a cyclic *interdependent* process, which can be *split only verbally*. The independent elements are fictitious and, structurally, have little or nothing to do with actual facts. The human nervous system represents, structurally, a mutually interdependent cyclic chain, where each partial function is in the functional chain, together with enforcing and 'inhibiting', and other mechanisms.

Up to this stage, we have used the term 'cyclic order', but, in reality, the order is *recurrent*, though of a character better described by the 'spiral theory', as explained in my *Manhood of Humanity* on p. 233. In the 'spiral theory', we find the foundation for this peculiar stratification in levels and orders, which is necessitated by the structure and function of the human nervous system. It should be noticed that the equations of the circle and spiral are non-linear, non-plus equations.

The above relation underlies a fundamental mechanism, known in psychiatry as 'sublimation', in which, and by which, quite primitive impulses, without losing their intensity and fundamental character, quite often- are transformed from very primitive levels, which frequently represent vicious and anti-social effects, into desirable characteristics, socially useful. Thus, a sadistic impulse may be sublimated into the socially useful vocation of the butcher, or, still further, into the skill and devotion to the service of their fellowmen, shown by many surgeons. We see that this mechanism is of tremendous importance, and responsible for what we call 'culture' and many other values. Our educational methods should understand this mechanism and apply this knowledge in the semantic training of youth. It is important to realize that this mechanism appears as the only semantic mechanism of correction which is in accord with the structure of the human nervous system, and so it seems workable. Various metaphysical preachings usually start by disorganizing the proper survival working of the human nervous system, and then we wonder that they fail, and that we cannot change 'human nature'. To deal with 'human nature', which is not something static and absolute, we need to approach it with more structural understanding and less prejudice. Then, and then only, can we eventually look for better semantic results.

The writer does not want the reader to conclude that, because in mathematics we have followed the survival order through extension, the mathematicians must, by necessity, be the sanest of the sane. Quite often this is not true, since many complexities exist which will be taken under analysis later.

Concluding remarks on order

One thing remains fundamental; namely, that the problems of order and extension are of paramount structural importance for sanity and our lives. They should be worked out and applied to the semantic training of the young in elementary education,. This would certainly produce a new generation saner than we are, and one which would, perhaps, lead lives less troubled than our own, and so, perhaps, of better survival value.

In concluding, it must be mentioned that a theory of *sanity*, because of the survival value of *order*, cannot *start* with the older, undifferentiated similarities, which are a product of *higher* abstractions, and thus of later origin, but *must* start with *differences* as fundamental, and so preserve the structure and order of the survival trend as applied in this work.

Animals do not possess such a highly differentiated nervous system as human beings. The difference between their higher and lower abstractions is thus not so fundamental, as we shall see later on. With them the question of *order* is less important, as they cannot alter it. Animals have the benefit of better co-ordination, since in them the above-described structural difficulties do not arise. They have normally no 'insane'. But, also, for the same reason, animals are not able to start every generation where the older left off. In other words, animals are not time-binders.

The structural complexity and differentiation of the nervous system in man is responsible, as is well known, not only for all our achievements and control over the world around us, but also for practically all our human, mostly semantic difficulties, many 'mental' ills included. The analysis in terms of *order* on the macroscopic level (semantic manifestations) reveals a profound connection with sub-microscopic processes of distribution., of nervous energy. When the mechanism which controls these processes is properly understood, then they can be controlled and educated by special semantic training. In other words, theoretical, doctrinal, higher abstractions may have a stabilizing and regulating *physiological* effect on the function of our nervous system.

The reader may be interested to know that 'order' is very important in animal life. An analysis of nest-building and the rearing of young among birds shows that each step of the cycle is necessary before the next step is taken. If the cycle is broken, they usually cannot adjust themselves to the new state of affairs, but must start from the beginning. This is a situation similar to our own when we cannot recall a line in a poem, but have to start from the beginning of the piece in order to recapture it. Pavlov was able, by the change of four-dimensional order of stimuli, to induce profound nervous disturbances in the nervous systems of his dogs, .

It appears, also, that in mild cases of aphasia, which is a neurological disturbance of linguistic processes, with word-blindness, word-deafness., the notion of 'order' and 'relations' is often the first to be disturbed. In some cases, lower order abstractions are carried out successfully, but calculation, algebra, and other higher order abstractions, which require *ordered* chains, become impossible. The aphasic seems to have a general incapacity for grasping *relations*, realizing *ordered series*, or grasping their succession.

We see that the problems of *order* are somehow uniquely important; and so the investigation of the psycho-logics of mathematics, which is based on *order*, might give us means of at least partial control of different undesirable human semantic afflictions.

But, after all, we should not be surprised that it is so. The structure of nervous systems consists of *ordered* chains produced by the impact of external and internal stimuli in a four-dimensional space-time manifold, which have a spatial and also a temporal *order*. The introduction of the finite velocity of nerve currents, which, although known, was, as a rule, disregarded by all of us, introduces automatically our *ordering* in 'space' and 'time' and, therefore, in space-time.

That is why the old anatomical three-dimensional analogies are vicious and false to facts when generalized. For better or worse, we happen to live in a four-dimensional world, where 'space' and 'time' cannot be divided. Whoever does this splitting must introduce fictitious, non-survival entities and influences into his system, which is moulded by this actual world and unable to adjust itself to fictions.

It seems obvious that all these problems of 'adjustment' and 'non-adjustment', 'fictitious' or 'actual' worlds., are strictly connected with our *s.r* toward these problems, and so ultimately with some structural knowledge about them. But *attitudes* involve lower order abstractions, 'emotions', affective components, and other potent semantic factors which we have usually disregarded when dealing with science and with scientific problems and method. For adjustment, and, therefore, for *sanity*, we must take into account the neglected aspects of science, of mathematics, and of scientific method; namely, their semantic aspects. In this way we shall abandon that other prevalent structural fiction referred to at the beginning of the present chapter; namely, that science and mathematics have an *isolated* existence.

The above considerations of order lead to a formulation of a fundamental principle (a principle underlying the whole of the non-aristotelian system); namely, that organisms which represent *processes* must develop in a cert*ain natural survival four-dimensional order*, and that the *reversal* of that order must lead *to pathological* (non-survival) developments. Observations disclose that, in all human difficulties, 'mental' ills included, a *reversal of the natural order* can be found as a matter of fact, once we decide to consider order as fundamental. Any identification of inherently different levels, or confusion of orders of abstractions, leads automatically to the reversal of natural order. As a method of preventive education and psychotherapy, whenever we succeed in reversing the reversed order or restoring the natural survival order, serious beneficial results are to be expected. These theoretical conclusions have been fully justified by experience and the work of Doctor Philip S. Graven in psychotherapy. It should be noticed that different primitive 'magic of words', or modern 'hypostatizations', 'reifications', 'misplaced concreteness', 'objectifications'., and all semantic disturbances represent nothing else but a confusion of orders of abstractions, or identifications in value of essentially different orders of abstractions.

Identification, or the confusion of orders of abstractions, in an *aristotelian* or *infantile* system, plays a much more pernicious role than the present official psychiatry recognizes. *Any* identification, at *any* level, or of *any* orders, represents a non-survival *s.r* which leads invariably to the reversal of the natural survival order, and becomes the foundation for *general* improper evaluation, and, therefore, *general* lack of adjustment, no matter whether the maladjustment is subtle as in daily life, or whether it is aggravated as in cases of schizophrenia. A *non-aristotelian system*, by a complete elimination of 'identity' and identification, supplies simple yet effective means for the elimination by preventive education of this general source of maladjustment. Book II is entirely devoted to this subject.

7. ON RELATIONS

One of the fundamental structural defects and insufficiencies of the traditional [A]-system was that it had no place for 'relations', since it assumed that everything could be expressed in a subject-predicate form. As we shall see, this is not true. Restriction to the subject-predicate form leaves out some of the most important structural means we have for representing this world and ourselves and has resulted in a general state of un-sanity. The explicit introduction of 'relations' is rather a recent innovation. A few words may be said about them, although the term 'relation' is one of the terms that we may accept as undefined, or that we may define in terms of multi-dimensional order.

Some relations, when they hold between A and B. hold also between B and A. Such relations are called *symmetrical*. For instance, the relation 'spouse'. If it holds between A and B, it holds also between B and A. If A is the spouse of B, B is the spouse of A. Terms like 'similarity' and 'dissimilarity' also designate relations of this kind. If A is similar or dissimilar to B, so is B similar or dissimilar to A. In general, a symmetrical relation is such that, if it holds between A and B, it also holds between B and A. In other words, the *order* in which we consider the relation of our entities is immaterial.

It is easy to see that not all relations are of such a character. For instance, in the relation 'A is the brother of B', B is not necessarily a brother of A, because B might be the sister of A. In general, relations which hold between A and B, but not necessarily between B and A, are called *non-symmetrical*. In these relations *order* becomes important. It is not a matter of indifference in what *order* we consider our entities.

If a relation is such that, if it holds between A and B. it *never* holds between B and A, it is called *asymmetrical*. Let us take, for instance, the relations 'father', 'mother', 'husband', . We readily see that if A is a father, or mother, or husband of B, B is *never* a father, or mother, or husband of A. The reversal of *order* is impossible in asymmetrical relations, and so any asymmetrical relation establishes a definite order.

Relations such as *before, after, greater, more, less, above, to the right, to the left, part,* and *whole,* and a great many others of the most important terms we have, are asymmetrical. The reader may easily verify this for himself. For instance, if A is *more* than B, B is *never* more than A, . We see at once that the troublesome little words, which are necessary to express *order* as 'before' and 'after'; terms of *evaluation*, such as 'more' and 'less'; and terms on which elementalism or non-elementalism depends, such as 'part' and 'whole', are in the list of asymmetrical relations.

Relations can be classified in another way, when three or more terms are considered. Some relations, called *transitive*, are such that, whenever they hold between A and B and also between B and C, they hold between A and C. For example, if A is before, or after, or above, or more., than B, and B is before, or after, or above, or more., than C, then A is before, or after, or above, or more., than C.

It should be noted that all relations which give origin to series are transitive. But so are many others. In the above examples, the relations were transitive and asymmetrical, but there are numerous relations which are transitive and symmetrical. Among these are relations of equality, of being equally numerous, .

Relations which are not transitive are called non-transitive. For instance, dissimilarity is not transitive. If A is dissimilar to B and B dissimilar to C, it does not follow that A is dissimilar to C.

Relations which, whenever they hold between A and B, and between B and C, *never* hold between A and C are called *intransitive*. 'Father', 'one inch longer', 'one year later'., are intransitive relations.

Relations are classified in several other ways; but, for our purpose, the above will be sufficient.

It is necessary now to compare the relational forms with the subject-predicate form of representation, which structurally underlies the traditional [A]-system and two-valued 'logic'. The structural question arises whether all relations can be reduced to the subject-predicate forms of language.

Symmetrical relations, which hold between B and A whenever they hold between A and B. seem plausibly expressed in the subject-predicate language. A symmetrical and transitive relation, such as that of 'equality', could be expressed as the possession of a common 'property'. A non-transitive relation, such as that of 'inequality', could also be considered as representing 'different properties'. But when we analyse *asymmetrical* relations, the situation becomes obviously different, and we find it a structural impossibility to give an adequate representation in terms of 'properties' and subject-predicates.

This fact has very serious semantic consequences, for we have already seen that some of the most important relations we know at present belong to the asymmetrical class. For example, the term 'greater' obviously differs from the term 'unequal', and 'father' from the term 'relative'. If two things are said to be unequal, this statement conveys that they differ in the magnitude of some 'property' without designating the greater. We could also say that they have different magnitudes, because inequality is a symmetrical relation; but if we were to say that a thing is unequal to another, or that the two have different magnitudes, when one of them was greater than the other, we simply should *not give an adequate account of the structural facts at hand*. If A is greater than B, and we merely state that they are unequal or of different magnitudes, we *imply the possibility* that B is greater than A, *which is false to facts*. To give an adequate account, and to prevent *false implications*, there is no other way than to say which one is greater than the other. We see that it is impossible to give an [A] *adequate* account when asymmetrical relations are present. The possession of the 'same' 'property', or of different 'properties', are both *symmetrical relations* and seem covered by the subject-predicate form. But it is impossible to account adequately for asymmetrical relations in terms of 'properties'. In other words, we see that a language and 'logic' based upon subject-predicate structure may perhaps express symmetrical relations, but fail to express adequately asymmetrical relations, because both 'sameness' and difference of predicates are symmetrical. Asymmetrical relations introduce a language of *new structure*, involving new *s.r.* Yet asymmetrical relations include many of the most important ones. They are involved in all *order*, all *series*, all *function*, in 'space', in 'time', in 'greater' and 'less', 'more' and 'less', 'whole' and 'part', 'infinity', 'space-time', . If we are restricted to the use of forms of representation unfitted for the expression of asymmetrical relations, ordinal, serial, functional, and structural problems could not be dealt with adequately. We should also have many insoluble semantic puzzles in connection with 'space', 'time', 'cause and effect', and many other relations in the world around-us, and ourselves.

A very interesting structural and semantic fact should be noticed that in symmetrical relations *order* is immaterial, in non-symmetrical relations it is important, and in asymmetrical relations *order* plays an all-important role and cannot be reversed. Order itself is expressed in terms of asymmetrical relations; as, for instance, 'before' or 'after', which apply to 'space', to 'time', 'space-time', 'structure'., and also to *all processes* and activities, the activities of the nervous system included. The asymmetrical relations 'greater', 'father'., imply ordering, while the 'unequal' (having different 'properties') or a 'relative'., do *not* imply ordering. If we consider subject-predicate forms as expressing a relation between the 'observer' and the 'observed', excluding humans, this last relation is also asymmetrical. Applying correct symbolism: if a leaf appears green to me, I certainly do not 'appear green' to the leaf! The last remark suggests that any [A] revision of the [A]-system is structurally impossible. To attempt a revision, we must begin with the formulation of a [non-A]-system of different structure.

The above simple considerations have very far-reaching consequences, as without relations, and particularly without asymmetrical relations, we cannot have *order*, and without order, in the analysis of processes, we are bound to introduce explicitly or implicitly some objectively meaningless 'infinite velocities' of the propagation of the process. Thus, the 'infinite velocity' of light, which is known to be false to facts, is at the very foundation of the [N]-system. The equally false to facts silent assumption of the 'infinite velocity' of nerve currents underlies [A] animalistic 'psychology, and results in elementalism. This *el* 'psychology', until this day, vitiates all human concerns and even all science, the newer quantum theories not excluded.

General non-elementalism and, in particular, its restricted aspect, the 'organism-as-a-whole', implies the relation of the 'parts' to the 'whole', for which we need asymmetrical relations. In the statement '*more* than an algebraic sum', 'more' is also an asymmetrical relation. When we analysed the statement, 'Smith kicks Brown', we saw that the problems of 'space', 'time', 'infinity'., entered, the solution of which requires *serial* notions, which evade analysis without asymmetrical relations.

The solution of the problems of 'space' and 'time' are fundamental for a theory of sanity, as they are potent structural factors in all *s.r.* In the majority of 'mentally' ill, we find a disorientation as to 'space' and 'time'. Similar milder forms of disorientation appear in all forms of semantic disturbances, as they are disturbances of evaluation and meanings in the form of delusional 'absolute space' and 'absolute time'. These semantic disturbances can be eliminated only by considerations of multi-dimensional order, which are impossible without asymmetrical relations, and so could not have been accomplished in an [A]-system.

The problems of multi-dimensional order and asymmetrical relations are strictly interdependent and are the foundation of structure and so of human 'knowledge'; and they underlie the problems of human adjustment and sanity. Without going into details, I shall suggest some relational and ordinal aspects as found in the structure and function of the human nervous system and their bearing on semantic reactions and sanity. I shall also apply these considerations to the analysis of a historically very important delusional factor which has influenced, until now, the *s.r* of mankind away from sanity. I am dealing only with selected topics, important for my purpose, which, to the reader, may appear one-sided and unduly isolated. In fact, all issues involved are strictly interconnected in a circular way, and no verbal analysis of objective levels can ever be 'complete' or 'exhaustive', and this should be remembered. On the [A] silent assumption of the infinite velocity of nervous impulses, that the nervous impulses spread 'instantaneously', 'in no time' (to use an Alice-in-Wonderland expression), order was of no importance. But when we take into account the *finite* and known velocity of nervous impulses, and the *serial*, chain structure of the nervous system, order becomes paramount. In such a serial structure, the problems of resistance, 'inhibitions', blockage, activation., become intelligible, so that some sane orientation is possible in this maze. It may be added that the intensity and the transformation of nervous impulses must somehow be connected with the paths they travel and are, therefore, problems to be spoken about in terms of order.

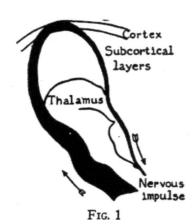

FIG. 1

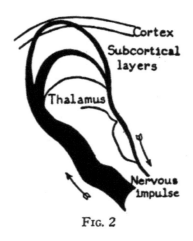

FIG. 2

What has just-been said may be illustrated by a rough and oversimplified hypothetical diagram. Fig. 1 shows how the normal (survival in man) impulse should travel. It should pass the thalamus, pass the sub-cortical layers, reach the cortex, and return. That the impulse is *altered* in passing this complicated chain is indicated in the diagram by the arbitrarily diminishing thickness of the line of the impulse.

Fig. 2 illustrates an hypothetical abnormal (non-survival in man) impulse. It emerges from the lower centres. For some nervous reason or other, the main impulse is blocked semantically, or otherwise, and does not reach the cortex; only a weak impulse does. What should be expected in such a case? We should expect regression to the level of activities of organisms which have no cortex, or a cortex very little developed. But this could not be entirely true, as organisms without a cortex have a nervous system adequate for their lives, activities., in their environment, with survival values. But a higher organism with a cortex, no matter how rudimentary, has the other parts of the nervous structure quite different in function, and without the cortex they are *inadequate* for survival, as experience shows. We see that the *order* in which the impulses pass, or are deviated from their survival path, is paramount. A great many different reasons may produce such deviation, too many to list conveniently. A great many of them are known, in spite of the fact that, in general, we know very little about nerve mechanisms. Suffice to say, that we know, on colloidal grounds and from experience, that macroscopic or microscopic lesions, drugs, and *false doctrines* affecting the sub-microscopic levels, may often produce similar end-results. Here I use the term 'false doctrines' in the *non-el* sense, and, therefore, take into account affective and *evaluation*-components, which are usually disregarded when we speak about 'false doctrines'.

Here we must consider a problem of crucial, general human significance. It seems evident that *evaluation* in life, and particularly in human life, represents a most fundamental psychological process underlying motivation and, in general, s.r, which shape our behaviour and result in collective structures which we may call 'stages of civilization'.

We may distinguish three periods of human development as characterized by their standards of evaluation:

1) The pre-human and primitive period of literal, general, and unrestricted identification. The semantics of this period could be formulated roughly as 'everything is everything else', which might be called one-valued semantics.

2) The infantile, or [A] period of partial or restricted identification, allowing symmetrical relations, to the exclusion of asymmetrical relations. Its semantics involve, among others, the 'law of identity'—'everything is identical with itself', its two-valued character being expressed by the postulate 'A is B or not B'.

3) The adult, or [non-A], or scientific period based on the complete elimination of identification, by means of asymmetrical and other relations, which establishes *structure* as the foundation of all 'knowledge'. Its semantics follow the ∞-valued semantics of probability and recognize 'equality', 'equivalence'., but no 'identity'.

Before analysing the above three periods separately, it must-be stated that 'identity', defined as 'absolute sameness', necessitates 'absolute sameness' in 'all' aspects, never to be found in this world, nor in our heads. Anything with which we deal on the objective levels represents a process, different all the 'time', no matter how slow or fast the process might be; therefore, a principle or a premise that 'everything is identical with itself' is *invariably false to facts*. From a structural point of view, it represents a foundation for a linguistic system non-similar in structure to the world or ourselves. All world pictures, speculations and *s.r* based on such premises must build for us delusional worlds, and an optimum adjustment to an *actual* world, so fundamentally different from our fancies, must, in principle, be impossible.

If we take even a symbolic expression $1 = 1$, 'absolute sameness' in 'all' aspects is equally impossible, although we may use in this connection terms such as 'equal', 'equivalent', . 'Absolute sameness in all aspects' would necessitate an *identity* of different nervous systems which produce and use these symbols, an *identity* of the different states of the nervous system of the person who wrote the above two symbols, an identity of the surfaces., of different parts of the paper, in the distribution of ink, and what not. To demand such impossible conditions is, of course, absurd, but it is equally absurd and very harmful for sanity and civilization to preserve until this day such delusional formulations as *standards of evaluation*, and then spend a lifetime of suffering and toil to evade the consequences. This may be comparable to the spending of many years in teaching and training our children that one and one *never* equal two, that twice two *never* equal four., and then they would have to spend a lifetime full of surprises and disappointments, if not tragedies, to learn, when they are about to die, that the above statements are always correct in mathematics and very often true in daily life, and finally acquire the sadly belated wisdom that they were taught false doctrines and trained in delusional *s.r* from the beginning.

If we revised these false doctrines, we would not twist the lives of younger generations to begin with. It seems that, for the sake of sanity, the term 'identity', symbolizing such a fundamental false structural doctrine, should be entirely eliminated from the vocabulary, but the term 'identification' should be retained in psychiatry as a label for extremely wide-spread delusional states which, at present, in a mild form, affect the majority of us.

If we investigate the standards of evaluation of animals, the experiments of Pavlov and his followers show that, after establishing a 'conditional reflex' (which means a physiological relating of a signal with food, for instance), the *physiological* effect of the signal on the nervous system of the animal is to produce secretions similar in quantity and quality to those the food produces. We can thus say that, from a physiological point of view, the animal organism *identifies* the signal with food. That represents the *animal standard of evaluation* at that given period. But even the animal nervous system is flexible enough to learn by experience that identification has no survival value, for, if, after the signal, food is repeatedly not forthcoming, he identifies again the signal with the absence of food. In more complex experiments, when both these identifications are interplayed, the result is a real physiological dilemma, culminating, usually, in a more or less profound nervous disturbance, corresponding to 'mental' ills in humans.

Identification represents a comparatively unflexible, rigid form of adaptation, of low degree conditionality, so to say, and, by neurological necessity, represents the processes of *animal* adaptation, inadequate for modern man. On human levels, it is found best exemplified in primitive peoples and in cases of 'mentally' ill. In less severe cases of semantic disturbances, we also find identification of different degrees of intensity. The milder cases are usually considered as 'normal', which, in principle, is very harmful, because it establishes an animalistic, or primitive, standard of evaluation for 'normal'. 'Identity', as we have seen, is invariably false to facts; and so identification produces, and must produce, non-survival *s.r*, and, therefore, must be considered *pathological* for modern man.

That identification afflicts the majority of us today is also shown by experiments with conditional reflexes and the psychogalvanic experiments which show clearly that the majority of humans *identify* the symbol with actualities, and *secretions very often follow*. In other words, the reactions are of such a low order of conditionality as we find in animals and in primitive men. In principle, it makes no difference whether a sound (or word), or other signal (symbol) is identified with food or other actualities which are not symbols, and the secretions are produced by the adrenal glands, for instance, resulting in fear or anger, instead of by the salivary or sweat glands. In all such cases, *in experiments* with humans, the evaluation is false to facts, and the *physiological secretion* is uncalled for if the *evaluation would be appropriate* to the situation. In very few instances, the human experiments with conditional and psychogalvanic reflexes break down, in the sense that the signal-symbol is *not identified* with first order actualities, and so such an organism has no uncontrolled glandular secretions for signal-symbols *alone*. In a [non-A]-system of evaluation, which involves on semantic levels the consciousness of abstracting, these exceptional persons (1933), with proper evaluation and controlled reactions, *prove the rule* for modern man. In other words, modern man, when he stops the pre-human and primitive identification, will have a much-increased and *conscious control* of his secretions, colloidal states of his nervous system., and so of his reactions and behaviour. The above applies to all *s.r*, 'logical' processes included.

Identification is found in all known forms of 'mental' ills. A symbol, in any form, or any *s.r* may be identified in value with some fictitious 'reality' at a given date, resulting in macro-physiological (glandular, for instance) or micro-physiological (colloidal.,) activities or disturbances which result in particular semantic states and behaviour. It is impossible to deny that 'mentally' ill have inappropriate standards of evaluation, and that identification appears always as an important factor in pathological evaluations. Experiments with 'mentally' ill show clearly that this evaluation can be altered or improved by different chemical agencies which affect the colloids of the nervous system, by environmental changes., and by *changing the standards of evaluation* which, at present, is usually called 'psychotherapy'. The analysis of the mechanism of evaluation leads, naturally, to a generalized and *simplified* method, which may have not only a therapeutic but also an important new *preventive value*.

Literal identification is found in all primitive peoples and accounts for their semantic states, reactions, their metaphysics, low development., but it is impossible, for lack of space, to go into details here.

The [A] standard of evaluation departed from literal identification to some extent. We still preserve in our school books as the most fundamental 'law of thought'—the 'law of identity'—often expressed in the form 'everything is identical with itself', which, as we have seen, is invariably false to facts. We do not realize that, in a human world, we are dealing at most only with 'equality', 'equivalence'., at a given place and date, or by definition, but never with 'identity', or 'absolute sameness', disregarding entirely space-time relations, involving 'all' the indefinitely many aspects which, through human ingenuity, we often manufacture at will. In an actual world of four-dimensional processes and the indefinitely many 'aspects' manufactured by ourselves, adjustment in principle is impossible, or, at best, only accidental, if we retain 'identity'. The [A] evaluation was based on symmetrical relations of 'identity' and also partial 'identity', expressed even in our political, economic., doctrines and corresponding behaviour, the analysis of which would require a special volume to be written, I hope soon, by some one.

Under the pre-human and primitive standards of evaluation, science was not possible. Under the [A] standards the beginnings of science became possible, but if science had not departed from those standards, we would have had no modern science. Lately, when the persecution of science has increasingly relaxed (not in all countries in a similar degree) and scientists were allowed to develop their disciplines with much less fear of persecution, sometimes even encouraged and helped by public interest, scientists found that they invariably had to build their own vocabularies of a distinctly, although unrealized, [non-A] character. The chasm between human affairs and science became wider and wider. The reason for it was that, in life, even at present, we preserve [A] standards of evaluation, and science mainly depends on subtler [non-A] means involving asymmetrical relations which alone can give us *structure*. I will return repeatedly, later on, to the [non-A] re-evaluation of the [A] standards of values.

The [non-A] evaluation is based on asymmetrical and other relations. I shall not attempt to summarize it here because the problems are very large and this whole volume is devoted to that subject. Here I shall mention, once more, that only with [non-A] standards of evaluation does a scientific treatment of man and his affairs become possible. A [non-A]-system depends on a complete elimination of identification which affects beneficially all our *s.r*, as experience and experiments show.

It has been already emphasized that in the human child the nervous system is not physically finished at birth, and that for some years thereafter it is plastic. Hence, the 'environment'—which includes languages, doctrines, with their structure, all connected with evaluation-components — conditions the future functioning of the system. The way in which the nervous system, works, the 'sanity', 'un-sanity' and 'insanity' of the individual depends to a large extent on how this plastic and sensitive apparatus is treated, particularly in childhood. Because of the *serial structure* of the nervous system, the language and doctrines supplied should be of the structure necessary for the adequate representation of serial structures and functions. With the old [A] means this could not be accomplished.

At this point, it will be well to introduce an important semantic subject, to which we shall return later; namely, the connection between the primitive subject-predicate language and identification. For example, the statement, 'the leaf is green', is taken to imply 'greenness', which, by its verbal structure, has the character of a 'substantive' and implies some sort of objective independence. It is *not* considered as an *asymmetrical relation* between the observer and the observed and, accordingly, tends toward an *additive* implication. 'Greenness' is thus objectified and *added* to the leaf in describing a 'green leaf'. The objectified 'greenness' leads to an anthropomorphic mythology, which, in turn, involves and develops the undifferentiated projecting mechanism so fundamental in semantic disturbances. The objectification is evaluated structurally as a 'real' situation, and this introduces the non-survival *reversed* order evaluation in which the use of the 'is' of identity, resulting in identification, is the main factor. The stronger the structural 'belief' in the 'truth' of the representation, or, in other words, the more we identify the higher order abstractions with the lower, which, in fact, are different, the more dangerous becomes the 'emotional' tension in the form of unjustified *evaluation*, which, ultimately, must involve *delusional* factors, no matter how slight, and result in semantic disturbances. *Ignorance*, involving strong faith in the erroneous structural belief, is dangerously akin to more developed symptoms of 'mental' illness called illusions, delusions, and hallucinations. We are mostly semantic victims of the primitive doctrines which underlie the [A] structure of our language, and so we populate the world around us with semantic phantoms which add to our fears and worries, or which lead to abnormal cheerfulness, well known among some 'mentally' ill.

It should be realized that in the [A] system of evaluation many individuals profit in various ways by what amounts to distracting the attention of mankind from actual life problems, which make us forget or disregard actualities. They often supply us with phantom semantic structures, while they devote their attention to the control of actualities not seldom for their personal benefit. If one surveys the [A] situation impartially, one occasionally feels hopeless. But, no matter how we now conspire one against another, and thus, in the long run, against ourselves, the plain realization that the difficulty is found in the standards of evaluation, establishes the necessary preliminary step to the escape.

It is a well-known fact that, in a large proportion of 'mental' ills, we find a semantic flight from 'reality' (*m.o*) when their 'reality' becomes too hard to endure. It is not difficult to see that different mythologies, cults., often supply such structural semantic 'flights from reality'; and that those who actually help, or who are professionally or otherwise engaged in producing and promulgating such semantic flights, help mankind to be un-sane, to deal with phantoms, to create dream states, . There is no longer any excuse in the old animalistic law of supply and demand— that, because there is a demand for such flights, they should be supplied. That argument is not held to apply to those who peddle drugs or wood alcohol. The flights from reality always have the earmarks of 'mental' illness. Very often such actively engaged individuals are themselves ill to the point of hallucinations; they often 'hear voices', 'see visions', 'speak tongues', . Very often other morbid symptoms occur which are similar to those shown by the 'mentally' ill of the usual hospital types. It is not generally realized that, although the patient suffers intensely, he usually shows marked resistance to any attempt to relieve him of his semantic affliction. Only *after* he is relieved by semantic re-education does the patient realize how very *unhappy he was*.

The situation is very serious. There is a powerful well-organized system, with enormous wealth behind it, based on [A] and pre-aristotelian standards of evaluation which keeps mankind in delusional semantic states. Its members do their best, better than they know, to keep mankind *un-sane* in flights from 'reality', instead of helping to revise the [A] standards of evaluation and to reorganize the horrible 'realities', *all of our own making*, into realities less painful. The comparatively few psychiatrists are naturally not a match for such vast numbers of well organized men and women who, in their blissful ignorance, work in the opposite direction; and all of us pay the price.

The activities of these individuals often promulgate something similar to the well-known 'induced insanity'. Quite often paranoid or paranoiac and, more rarely, hypomanic patients can influence their immediate companions to such an extent that they join in believing in their delusions and copy their *s.r.* Susceptible associates begin to develop similar delusions and hallucinations and to pass through episodes themselves, perfectly oblivious to contradictions with external *m.o* reality. There are many paranoiac-like semantic epidemics of this kind on record. It is instructive to visit some 'meetings' and watch the performer and the audience. The pathetic side of it is that these performers, themselves not realizing the harmfulness of the situation, often pretend, or genuinely believe, that they are helping mankind by preaching some metaphysical 'morals'. What they *actually* produce is a disorganization of the survival-working of the human nervous system, particularly if they train the structurally undeveloped nervous system of children to delusional evaluations and *s.r*, and, in general, make sanity and higher and effective ethical standards very difficult or impossible. It is positively known that *s.r* are inextricably connected with electrical currents, secretions of different glands., which, in turn, exert a powerful influence on colloidal structure and behaviour, and so condition our neurological and physiological development. There can be no doubt that imposing delusional *s.r* on the undeveloped child must result in at least colloidal injury, which later on facilitates arrested development or regression, and, in general, leads away from adjustment and sanity.

Lack of space and the essentially constructive aims of the present system do not allow me to analyse many fundamental interrelations in the development of man, but a brief list, worthy of analysis, may be suggested:

1) The relation between the pre-human reactions and the reactions of the primitive man, involving always some *copying* by mutants of the responses of the prevailing simpler organisms.

2) The interrelation between the reactions of the primitive man, his animism, anthropomorphism, his other *s.r* and the *structure* of his language and semantics.

3) The relation between the structure of primitive languages and the structure of the 'philosophical grammar' formulated by Aristotle, generally called 'logic'.

4) The relation between this grammar, the structure of language, and the further development of our structural metaphysics and *s.r.*

5) The influence the last conditions exerted on the structure of our institutions, doctrines, and the *s.r* related to them.

6) The relation between the 'copying animals in our nervous processes' and semantic blockages., preventing an adult civilization, agreement, sanity, and other desirable human reactions.

This brief list suggests an enormous field for further research, but, even now, the formulation of a [non-A]-system of evaluations makes a few points dearer.

An infant, be it primitive or modern, begins life with *s.r* of identity and confusion of orders of abstractions, natural to his age, yet false in principle, and structurally false to fact. At present, parents and teachers seldom check or counteract this tendency, mostly not realizing the importance of this semantic factor and its role in the future adjustment of the individual. In the rough, to a baby, his cry 'is' food. Words 'are' magic. This identification is structurally false to facts, but in *babyhood it mostly works*. To the infant, experience proves that the noises he makes, a cry or a word, have the objective value,—food. The semantic identity of the symbol and the un-speakable object level,—food,—has been established. This infantile attitude or *s.r* is carried on into grown-up life.

Under very simple conditions of primitive peoples, in spite of many difficulties, this attitude of identification is not always checked by experience, and experimenting is non-existent at this stage. If it is, then such checking of identification is 'explained' by some sort of demonology and 'good' or 'evil' 'spirits', . Delusional, from the modern point of view, *s.r* are compensated by mythologies, making the two sides of the semantic equation equivalent. This equating tendency is *inherent* in all human *s.r.* It expresses the instinctive 'feel' for the similarity of structure as the base of 'knowledge', and it ultimately finds its expression in mathematical equations. In all psycho-logical processes of 'understanding', we must have some standards of evaluation and 'equivalence'. On primitive levels, this is accomplished by literal identification and delusional mythology of the type, that a storm at sea is 'caused' by a violent quarrel between a 'god' and his 'wife'; or, in contemporaneous mythology, a draught, or fire, or death by lightning, is explained as 'punishment' for 'sins', . Semantic compensation is needed and produced. A similar semantic process produces scientific theories, but with different standards of evaluation. At present, scientific theories do not cover all semantic needs and urges of mankind, owing to the prevailing false to fact identification of different orders of abstractions. With the full consciousness of abstracting, which means proper evaluation or differentiation between orders of abstractions, science will then cover all our non-pathological semantic needs, and different primitive mythologies will become unnecessary. A very harmful, primitive, delusional semantic factor of blockages would be eliminated.

The 'is' of identity plays a great havoc with our s.r, as any 'identity' is structurally false to fact. An infant does not know and cannot know that. In his life, the 'is' of identity plays an important semantic role, which, if not checked intelligently, becomes a pernicious semantic factor in his grown-up reactions, which preserve the infantile character and with which *adult* adjustment and semantic health is impossible. The infant begins to speak and again he is trained in the 'is' of identity. Symbols are identified with the un-speakable actions, events and objects under penalty of pain or even death. The magic of words begins its full sway. As a rule, parental, crude disciplining of the infant, particularly in former days, trained the s.r of the infant again in the delusional 'is' of identity. The results are semantically and structurally very far-reaching and are found to underlie modern mythologies, militarism, the prevailing economic and social systems, the control by fear (be it 'hell' or machine guns), illusory gold standards, hunger, .

Experience shows that such identification of symbols with the un-speakable levels works very well with animals. With man, it leads only to the misuse of the human nervous system, semantic disturbances of evaluation, and the prevailing animalistic systems in practically all fields, resulting in the general chaos in human affairs.

It should be noticed that the 'is' of predication also expresses a sort of *partial identity*, leading to primitive anthropomorphism and general confusion of orders of abstractions. By an inherent necessity, our lives are lived on the un-speakable objective levels, which include not only ordinary objects but also actions and immediate feelings, symbols being only auxiliary means. The natural ordinal evaluation, which should be the foundation for healthy s.r, appears as the event-process level first, the object next in importance; the objective level first, the symbolic next in importance; the descriptive level first, the inferential level next in importance, . The semantic *identification* of these different levels not only abolishes the natural evaluation, but, in fact, reverses the natural order. Once this is realized, we see clearly that all statements about the objective level, which is made up of absolute individuals, are only *probable* in different degrees and can never be certain. The 'is' of identity underlies, also, the two-valued, too primitive, too restricted, and structurally fallacious [A] 'logic'.

The crucial semantic importance of asymmetrical relations becomes obvious when we consider that all *evaluation* and *non-el* meanings depend ultimately on asymmetrical relations. In the technical fields, mathematics and the exact sciences; in the semi-scientific fields, economics, politics, sociology.; in the as yet non-scientific fields, 'ethics', 'happiness', 'adjustment'., represent ultimately different forms of *evaluation*, impossible to formulate adequately under aristotelianism.

Obviously, a [non-A]-system based on proper semantic evaluation leading to non-pathological reactions, adjustments., must make relations and multi-dimensional order fundamental for sanity. The semantic connection between mathematical methods and all the other concerns of man becomes also necessary and obvious.

In mathematics, recently, the notion of equality needed a refinement and the notion of 'identity' has been introduced. The present analysis discloses that, although the refinement and the symbol may be retained, yet the name should be entirely abandoned, because it conceals a very semantically vicious confusion of orders of abstractions. If, by definition, we produce new terms, these new terms are of a higher order abstraction than the terms used in the definition, and so the *identification* of them as to the orders of abstractions is *physiologically* and structurally false to facts.

The problems discussed in the present chapter have been felt vaguely for more than two thousand years and found their first historical expression in the rift between Aristotle, the biologist, and Plato, the founder of mathematical philosophy. Mathematics is, in principle, [non-A], and so, in the study of mathematics, we can learn most about the principles of non-aristotelianism. In physics, only very recently, do we begin to eliminate the 'is' of identity and elementalism which resulted in the [non-N] systems. All sciences strive to become more mathematical and exact and so [non-A]. In fact, all advances in science are due to the building of new [non-A] languages, usually called 'terminology'. We can go further and say, definitely, that, to have any science, we must make a [non-A] revision of the languages used. Similarly with 'man', either we decide to introduce into human affairs scientific evaluation, and so part company with the [A] and pre-aristotelian system of evaluation, or preserve [A] structure, and have no science of man, or science of sanity, but continue in the prevailing chaos.

8. ON THE NOTIONS OF 'MATTER', 'SPACE', 'TIME'

The facts at hand in 1933 show that the language we use for the purpose of describing events *is not* the events; the representation symbolizes what is going on inside our skins; the events are outside our skins and *structural similarity* is the only link between them. Historically, as a race, we learned sooner and more about the events outside our skins than about the events inside our skins; just as a fish or a dog 'knows' a lot about his world, lives sometimes happily and abundantly, and yet 'knows' nothing about biology, or physiology, or psycho-logics. Only recently did we begin to study ourselves scientifically. At some stage of our development, we introduced structurally simple forms of representation, such as a *language* of subject-predicate, of additivity, . We are still perplexed when we find that the events outside our skins cannot be pressed into schemes which are manufactured inside our skins. Our nervous system, with its ordered and cyclic structure and function, manufactures abstractions of different orders, which have quite distinct structure and different characteristics. On different levels, we manufacture different abstractions, dynamic and static, continuous and discontinuous., which have to take care of our needs. If the verbal schemes we invent do not fit structurally the world around us, we can always invent new schemes of new structure which will be more satisfactory. It is not a problem of the world around us, for our words cannot change that, but of *our ingenuity.* In the meantime, we learn something very important; namely, about the world's *structure,* which is the only content of knowledge.

There are good structural reasons why the world should, or should not, be accounted for in terms of differential equations, or in *terms and language* of 'causality', . The term *order* is structurally fundamental and will help us in a radical and constructive way, in our quest.

First, however, we will investigate some further semantic problems, remembering that a theory of sanity, which means a theory of adjustment, should emphasize the methodological and structural means for such semantic adjustment. The dynamic-static translations are fundamentally connected with different orders of abstractions and involve psycho-logical issues connected with 'emotions' and 'intellect', linearity versus non-linearity, 'straight' versus 'curved'., explained in Parts VII and VIII.

In life, as well as in science, we deal with different happenings, objects, and larger or smaller bits of materials. We have a habit of speaking about them in terms of 'matter'. Through a *semantic disturbance,* called identification, we fancy that such a thing as 'matter' has separate physical existence. It would probably be a shock to be invited seriously to *give* a piece of 'matter' (give and *not* burst into speech). I have had the most amusing experiences in this field. Most people, scientists included, hand over a pencil or something of this sort. But did they actually give 'matter'? What they gave *is not* to be *symbolized* simply 'matter'. The object, 'pencil', which they *handed,* requires linguistically 'space'; otherwise, there would be no pencil but a mathematical point, a fiction. It also requires verbally 'time'; otherwise, there would be no pencil but a 'flash'.

Similarly, if any one is invited to *give* a piece of 'space' (again *give* it, and *not* burst into speech), the best he could do would be to wave his hand and try to show 'space'. But the waving of the hand referred to what we call air, dust, microbes, gravitational and electromagnetic fields, . In other words, structurally, the supposed 'space' was *fulness* of some materials already 'in space' and 'in time'.

In the case of *giving* 'time', one could *show* his watch. A similar objection holds, also; namely, that he has shown us so-called 'matter' which is 'moving' in 'space'. It is very important to acquire the *s.r* that when we use the term 'matter' we refer to something, let us say, the pencil, which, according to the accepted *el language*, also involves 'space' and 'time', which we disregard. When we use the term 'space', we refer to a fulness of some materials, which exists in 'time'. But because these materials are usually invisible to the 'senses', we again disregard them. In using the term 'time', we refer to 'matter' moving in 'space', which again we disregard.

What is said here and what will follow is structurally unconditionally fundamental for a theory of sanity, because in most cases of 'insanity' and un-sanity, there is a disorientation as to 'space' and 'time'. In identification, the semantic disturbance which affects nearly all of us, and is at the foundation of the majority of human difficulties, private or public, there invariably appears a special semantic disorientation in our feelings toward 'matter', 'space', and 'time'. This is only natural, for the 'insane' and un-sane are the unadjusted; the 'sane' are the supposedly adjusted.

Adjusted to what? To the world around us and ourselves. Our *human* world differs from the world of animals. It is more complex and the problems of human adjustment become also more subtle. In animal life, attitudes toward the world do not matter in a similar sense; with us, they become important; hence the need of analysis of the new human 'semantic universe', which involves the 'universe of discourse'. This 'universe of discourse' is strictly connected with the *terms* of 'matter', 'space', and 'time', structure, and our semantic *attitude* toward these terms.

Let us return to the analysis of our object which we call the 'pencil'. We have seen that the *object* pencil *is not* 'matter', nor 'space', nor 'time'. A question arises, which has been asked very often and has *never*, to my knowledge, been answered satisfactorily: what 'is' the *object* pencil, and what 'are' the *terms* 'matter', 'space', and 'time'? Here and there some one has given fragments of answers or some satisfactory detached statements. But in every case I know, the semantic disturbance called identification appears, and so even the casual correct answer is not applied but remains enmeshed in some other identifications. I have spent much 'time' and labour in overcoming my own identifications, and now confront the situation that nearly every work I read from this point of view cannot be criticized, but requires rewriting. This task is impossible for me, technically and otherwise. So, finally, I decided to formulate the present [non-A]-system, and then see what kind of reconstruction can be accomplished with the new evaluation.

The answer to the questions set above is childishly simple, yet I will carry it all through and let the semantic consequences speak for themselves. The chunk of nature, the specially shaped accumulation of materials., which we call a pencil, 'is' fundamentally and *absolutely un-speakable*, simply because whatever we may *say* about it, *is not it*. We may write with this something, but we cannot write with its name or the *descriptions* of this something. So the object *is not words*. It is important that the reader be entirely convinced at this point, and it requires some training, performed repeatedly, before we get our *s.r* adjusted to this simple fact. Our statement had two parts. One was rather-unpromising; namely, that the object was absolutely un-speakable, because no amount of words will make the object. The other was more promising, for we learned an extremely important, perhaps crucial, semantic fact; namely, what the *object* pencil *is not*; namely, that the *object is not words*. We must face here an important semantic fact. If we are told that we cannot get the moon, we stop worrying about it, and we regard any dream about getting the moon as an infantile phantasy. In this example, we could not even say that such news as the impossibility of getting the moon was sad, or unpleasant, news. We might say so, jokingly, to an infant, but the majority of the grown-ups would not have their *s.r* perturbed by it. A similar situation arises with the object called pencil. The *object is not words*. There is nothing sad or depressing about this fact. We accept it as a fact and stop worrying about it, as an infant would. The majority of the old 'philosophical' speculations about this subject belong to the semantic period of our infancy, when we live in phantasies and structurally gamble on words to which we affectively ascribe objective existence, . This represents full-fledged un-sanity due to identification. The answer to the question, what 'are' the *terms* 'matter', 'space', and 'time', is, as usual, given in the properly formulated question. They 'are' *terms,—'Modi considerandi'*, as Leibnitz called them without fully realizing the semantic importance of his own statement. Incidentally, it must be noticed that it was the psycho-logical characteristic of Leibnitz who was capable of such a statement, that was probably responsible for his whole work, as will become more apparent later on. When we abandon primitive standards of evaluation, geniuses will be *made* by a semantic education which relieves the race from the older blockages.

So we see clearly that outside of our skins there is something going on, which *we call* the world, or a pencil, or anything, which is *independent* of our words and which is *not* words. Here we come across a very fundamental irreversible process. We can say that in this world a man and his words have happened. There is a 'causal' eventual complex series between the world, us, and our words, but in unaided nature this process is, in the main, irreversible, a fact unknown to primitives who believe in the magic of words. Through our ingenuity we can make this process partially reversible; namely, we can produce gramophones, telephones in all their developments, electromechanical men who obey orders, .

We know in 1933 that in the semantic world this process is dramatically effective. Words are the result of the activity of one organism, and they, in turn, activate other organisms. On the macroscopic level of ordinary behaviour, this last was known long ago, but only in the last few years has psychiatry discovered what kind of semantic and psychophysiological disasters words and their consequences may produce in the human organism. These last are already on sub-microscopic levels, not obvious and, therefore, only recently discovered.

The structure of the language of 'matter', 'space', and 'time' is ancient. The primitive saw something, ate something, was hurt by something, . Here was an occasion for a grunt of satisfaction, or of pain. The equivalents of words like 'matter', 'substance'., originated. Neither he, nor the majority of us, realized that the small or large bits of materials we deal with appear as extremely complex *processes* (explained in Part X). For him, as for most of us, these bits of materials 'are' 'concrete', whatever that means, and he might know 'all about them', which must have led to identification., and other delusional evaluations. Of course, these were phantasies of human infancy, and, in lives lived in a world of phantasies, adjustment, and, therefore, sanity, is impossible. Since he did not see or feel, or know about the material he was immersed in, the *fulness* he was living in, he invented the term 'space', or its equivalent, for the *invisible* materials which were present. Knowing nothing of fulness, he objectified what appeared to him to be empty space, into 'absolute emptiness', which later became 'absolute space', 'absolute nothingness', by 'definition'.

There are several important remarks which can be made about this 'absolute emptiness' and 'absolute nothingness'. First of all, we now know, theoretically and empirically, that such a thing does not exist. There may be more or less of something, but never an *unlimited* 'perfect vacuum'. In the second place, our nervous make-up, being in accord with experience, is such that 'absolute emptiness' requires 'outside walls'. The question at once arises, is the world 'finite' or 'infinite'? If we *say* 'finite', it *has* to have outside walls, and then the question arises: What is 'behind the walls'? If we say it is 'infinite', the problem of the psychological 'walls' is not eliminated, and we still have the semantic for walls, and then ask what is beyond the walls. So we see that such a world suspended in some sort of an 'absolute void' represents a *nature against human nature*, and so we had to invent something *supernatural* to account for such assumed nature against human nature. In the third place, and this remark is the most fundamental of all, because a symbol must stand for something to be a symbol at all, *'absolute nothingness' cannot be objective and cannot be symbolized at all.* This ends the argument, as all we may say about it is neither true nor false, but *non-sense.* We can make noises, but say nothing about the external world. It is easy to see that 'absolute nothingness' is a *label for a semantic disturbance*, for verbal objectification, for a pathological state inside our skin, for a fancy, but not a symbol, for a something which has *objective* existence outside our skin.

Some other imaginary consequences of this semantic disturbance are far-fetched and very gloomy. If our world and all other worlds (island universes) were somehow suspended in such an 'absolute void', these universes would radiate their energy into this 'infinite void', whatever that means, and so sooner or later would come to an end, their energy being exhausted. But, fortunately, when we eliminate this pathological semantic state by proper education all these gloomy symptoms vanish as mere fancies. It must be noticed that this 'absolute space', 'absolute void', 'absolute nothingness' with its difficulties, which are due to very primitive structural speculations on *words*, and to some un-sane ascribing of objectivity to words, can be abolished quite simply if we decide to investigate and re-educate our *s.r.*

What we know positively about 'space' is that it is not 'emptiness', but 'fulness', or a 'plenum, . Now 'fulness' or a 'plenum', first of all, is a term of entirely different *non-el* structure. When we have a plenum or fulness, it must be a plenum of 'something', 'somewhere', at 'some time', and so the *term implies*, at least, *all three of our former elementalistic terms.* Furthermore, fulness by some psycho-logical process does not require 'outside walls'. If we ask if such a universe of fulness is 'finite' or 'infinite', without any psycho-logical difficulties, we may reply that we do not know, but if-we study enough of the materials of this universe we *may know.* A universe of fulness may be assumed to have boundaries, and then we may ask again the annoying question: What is beyond? With the proper use of *language*, this difficulty is again eliminated.

Without going into unnecessary details, we see that a boundary, or a limit, or a wall, is something by *definition*, beyond which we cannot go. If there is nothing to restrict our progress, there are no *boundaries.* Let us fancy some cosmic traveller with some extraordinary flying machine, and let us assume that he flies without stopping in a 'definite direction'. If he never encounters any boundary, he is surely entitled to say that his universe is unbounded. The question may arise: Is such an unbounded universe finite or infinite in size? Again let us apply correct language and a little analogy. A traveller on a sphere, like our own earth, could travel *endlessly* without ever coming to a boundary, and yet we know that the sphere, our earth, is of finite size. Mathematicians have worked out this point, and it is embodied in the Einstein theory. The universe is unbounded, an answer which satisfies our feelings; yet it is finite in size, although very large, an answer which satisfies our rationality.* The visualization of such a universe is quite difficult. It should not be visualized as a sphere, but, at a later stage, we will see that it *can* be visualized satisfactorily. The condition for visualization is to eliminate identification, that *semantic disturbance* which is strictly connected with primitive ways of 'thinking'.

The problems of 'time' are similar, although they have a different neurological background. The rough materials we deal with mostly affect our sight, touch . Invisible materials, like air., affect these 'senses' less, but more the kinesthetic 'senses' by which muscular movements are appreciated, and so 'space' and 'time' have different neurological backgrounds. 'Time' seemingly represents a general characteristic of all nervous tissue (and, perhaps, living tissue in general)

*I do not introduce here the latest speculations in this field, because, from a non-aristotelian point of view, they appear meaningless.

connected with summarizing or integrating. What we have to deal with in this world and in ourselves appears as periods and periodicity, pulsations, . We are made up of very long chains of atomic pulsating clocks, on the submicroscopic level. On the macroscopic level, we have also to deal with periodic occurrences, of hunger, sleep, breathing, heart-beats, . We know already that, beyond some limits, discontinuous times, when rapid enough, are blended into continuous feelings of pressure, or warmth, or light,. On objective levels we deal with times, and we feel 'time', when the times are rapid enough.

Again, the moving pictures are a good illustration. The normal moving-picture film shows sixteen pictures a second. The film gives us static pictures with finite differences. When we put it on the projector, the differences vanish. Our nervous system has summarized and integrated them, and we see 'continuous motion'. If pictures are taken at the rate of eight a second and then run on the normal projector for the speed of sixteen a second, we summarize and integrate again, but we see a fast moving picture. If the pictures are taken at the rate of 128 exposures a second and run on the normal projector of sixteen pictures to a second, we have what is called a slow moving picture. It should be noticed that the order of the semantic rhythmic processes is fourfold; it involves order not only in 'space' (three dimensions) but in 'time' also. Periods of contraction alternate with periods of rest, and this occurs at nearly regular intervals.

This rhythmic tendency is, indeed, so fundamental and so inherent in living tissue that we can, at pleasure, make voluntary muscles; for example, exhibit artificially induced rhythmic contractions by immersing them in special saline solutions, as, for instance, a solution of sodium chloride. We should also not wonder why modern science assumes that life may have originated in the sea. The physico-chemical conditions of saline solutions are such that they favour rhythmic processes; they not only may originate them, but may also keep them up, and life seemingly is very closely connected with autonomous rhythmic processes.

Such rhythmic processes are *felt* on lower orders of abstraction as 'continuous time', probably because of the rapidity and overlapping of periods. On higher order abstractions, when structurally proper linguistic and extra-neural means are developed, they appear as times.

Perhaps, neurologically, animals *feel* similarly as we do about 'time', but they have no neurological means to elaborate linguistic and extra-neural means which alone allow us to extend and summarize the manifold experience of many generations (time-binding). They cannot pass from 'time' to 'times'. Obviously, if we do not, we then renounce our human characteristics, and copy animals in our evaluating processes, a practice which must be harmful.

In nature the visible and invisible materials seemingly consist of recurring pulsations of extremely minute and rapid periods, which, in some instances, become macroscopic periods. In the first case, we cannot see them or feel them, so we talk about 'concreteness', . In the second case, we see the periodic movements, as of the earth around the sun., or we feel our heart-beats, . We see that the visible or invisible materials in nature are compounded of periodic pulsations and are simply two aspects of one process. The splitting of these processes into 'matter', 'space', and 'time' is a characteristic function of our nervous system. These abstractions are *inside* our skins, and are methods of representation for ourselves to ourselves, and *are not* the objective world around us.

It must be realized that under such circumstances we cannot speak about 'finiteness' or 'infiniteness' of 'matter', 'space', and 'time', as all the old 'philosophers' have done, Leitnitz included, because these terms 'finite' and 'infinite', though they may be conceivably applied to *numbers* of aspects of objective entities, have *no meaning* if applied to linguistic issues, that is, to *forms of representation* outside of numbers. Of course, if, through a semantic pathological disturbance (objectification), we do ascribe some delusional objective existence to verbal terms, we can then talk about anything, but such conversations have no more value than the deliria of the 'mentally' ill. The terms 'finite' or 'infinite' are only legitimately applied to *numerical* problems, and so we can speak legitimately of a finite or infinite numbers of inches, or pounds, or hours, or similar entities, but statements about the 'finite mind' or the 'understanding of the infinite'., have no meanings and only reveal the pathological semantic disturbance of the patient.

The objectification of *our feeling* of 'time' has had, and has at present, very tragic consequences strictly connected with our un-sanity. It must be remembered that particularly in 'mental' and nervous difficulties the patient seldom realizes the character of his illness. He may feel pains, he may feel very unhappy, and what not, but he usually does not understand their origin. This is particularly true with semantic disturbances. One may explain endlessly, but, in most cases, it is perfectly hopeless to try to help. Only a very few benefit. Here lies, also, the main difficulty in writing this book. Readers who identify, that is, who believe unconsciously with all their affective impulses in the objectivity of 'matter', 'space', and 'time', will have difficulty in modifying their *s.r* in this field.

Let us see now what consequences the objectification of 'time' will have for us. If we do *not* objectify, and *feel* instinctively and permanently that words *are not* the things spoken about, then we could not speak about *such meaningless* subjects as the 'beginning' or the 'end' of 'time'. But, if we are semantically disturbed and objectify, then, of course, since objects have a beginning and an end, so also would 'time' have a 'beginning' and an 'end'. In such pathological fancies the universe must have a 'beginning in time' and so must have been made., and all of our old anthropomorphic and objectified mythologies follow, including the older theories of entropy in physics. But, if 'time' is only a *human form of representation* and *not an object,* the universe has no 'beginning in time' and no 'end in time'; in other words, the universe is 'time'-less. It was not made, it just 'was, is, and will be'. The moment we realize, feel permanently, and utilize these realizations and feelings that words *are not* things, then only do we acquire the semantic freedom to use different forms of representation. We can fit better their structure to the facts at hand, become better adjusted to these facts which *are not* words, and so evaluate properly *m.o* realities, which evaluation is important for sanity.

According to what we know in 1933, the universe is 'time'-less; in other words, there is no such *object* as 'time'. In terms of periods, or years, or minutes, or seconds, which is a *different language*, we may have infinite numbers of such times. This statement is another form of stating the principle of conservation of energy, or whatever other fundamental higher abstraction physicists will discover.

Because 'time' is a *feeling*, produced by conditions of this world outside and inside our skins, which can be said to represent times, the problem of 'time' becomes a neuro-mathematical issue. It must also be noticed that times, as a term, implies times of something, somewhere, and so, as with plenum or fulness, it is structurally a *non-el*, [non-A] term.

Times has also many other most important implications. It implies *numbers* of times, it implies periods, waves, vibrations, frequencies, units, quanta, discontinuities, and, indeed, the whole structural apparatus of modern science.

Euclidean 'space' had the semantic background of 'emptiness'. In it we moved our figures from place to place and always assumed that this could be done quite safely and accurately. Newtonian mechanics also followed this path and even postulated an 'absolute space' (emptiness). All of which harks back to the old aristotelianism.

[non-E], [non-N], and [non-A] systems have the semantic background of fulness or plenum, although, unfortunately, this background is, as yet, mainly unrealized, not fully utilized; it has not, as yet, generally affected our *s.r.*

It should be noticed that scientists, in general, disregard almost completely the verbal and semantic problems explained here, a fact which leads to great and unnecessary confusion, and makes modern works inaccessible to the layman. Take, for instance, the case of the 'curvature of space-time'. Mathematicians use this expression very often and, inside their skins, they know mostly what they are talking about. Millions upon millions of even intelligent readers hear such an expression as the 'curvature of space-time'. Owing to nursery mythology and primitive *s.r*, 'space' for them is 'emptiness', and so they try to understand the 'curvature of emptiness'. After severe pains, they come to a very true, yet, for them, hopeless, conclusion; namely, that 'curvature of emptiness' is either *non-sense* or 'beyond them', with the semantic result that either they have contempt for the mathematicians who deal with non-sense or feel hopeless about their own capacities—both undesirable semantic results.

The truth is that 'curvature of emptiness' has no meanings, no matter *who* might say it, but curvature of fulness is entirely different. Let the reader look at the cloud of smoke from his cigarette or cigar, and he will at once understand what 'curvature of fulness' means. Of course, he will realize, as well as the mathematicians do, that the problem may be difficult, but, at least, it *has sense* and represents a problem. It is not non-sense.

Similar remarks apply to higher dimensions in 'space'. Higher dimensions in 'emptiness' is also non-sense; and the layman is right in refusing to accept it. But higher dimensions in fulness is entirely a different problem. A look at the cloud of smoke from our cigarette will again make it completely plain to everybody that to give an account of fulness, we may need an enormous number of data or, as we say roughly, of dimensions. This applies, also, to the new four-dimensional world of Minkowski. It is a fulness made up of world lines, a network of events or intervals., and it is not non-sense.

For the reasons already given, I do not use the terms 'matter', 'space', or 'time' without quotation marks; and, wherever possible, shall use, instead, the terms 'materials', 'plenum', 'fulness', 'spread', and 'times', (say seconds). Indeed, these semantic problems are so serious that they should be brought to the attention of International Mathematical and Physical Congresses, so that a new and *structurally correct* terminology could be established. It is *not* desirable that science should *structurally mislead* the layman and disturb his *s.r.* It is easier for trained specialists to change their terminology than to re-educate semantically the rest of the race. I would suggest that terms 'matter', 'substance', 'space', and 'time' should be completely eliminated from science, because of their extremely wide-spread and vicious structural and so semantic implications, and that the terms 'events', 'space-time', 'material', 'plenum', 'fulness', 'spreads', 'times'., be used instead. These terms not only do not have the old structural and semantic implications, but, on the contrary, they convey the *modern* structural notions and involve new *s.r.* The use of the old terms drags in, unconsciously and automatically, the old primitive metaphysical structure and *s.r* which are entirely contradicted by experience and modern science. I venture to suggest that such a change in terminology would do more to render the newer works intelligible than scores of volumes of explanations using the old terminology.

Protoplasm, even in its simplest form, is sensitive to different mechanical and chemical stimulations; and, indeed, undifferentiated protoplasm has already all the potentialities of the future nervous system. If we take an undifferentiated bit of protoplasm, and some stimulation is applied to some point, the stimulus does not spread somehow 'all over at once', with some mysterious 'infinite' velocity, but propagates itself with finite velocity and a diminishing gradient from one end of our bit of protoplasm to the opposite end.

Because of the *finite velocity* of propagation and the fact that the *action is by contact in a plenum*, the impulse has a definite direction and diminishing intensity, or, as we say, the bit of protoplasm acquires a temporal polarity (head-end). Such polarity conditions produce a directed wave of excitement of diminishing intensity, which Child calls a dynamic gradient. If such a stimulation were applied to one spot for a considerable length of 'time', some kind of polarization may become lasting. In some such way those dynamic gradients have become *structuralized* in the forms of our nervous system, which represent the preferred paths by which the nervous impulses travel.

The bodies of most organisms are protected from outside stimulation by some kind of membrane or cuticle and the parts of the surface have developed so as to be sensitive to one form of stimulation and not to others. For instance, the eye registers the stimulations of light waves, while it is insensitive to sound., and, even if hit, it gives only the feeling of light. Each 'sense-organ' has also the nervous means of concentrating stimuli, intensifying them., and so of effecting the most efficient response of the corresponding end-organ.

Is the problem of adjustment in the animal world similar to that in the human world? No, it is entirely different. Animals do not alter their environment so rapidly, nor to such an extent as humans do. Animals are not time-binders; they have not the capacity by which each generation can start where the former left off. Neurologically, animals have no means for extra-neural extensions, which extensions involve the complex mechanism with which we are dealing throughout this work.

The example of the caterpillar, already cited, shows clearly how organisms not adapted to their environment perish and do not propagate their special, non-survival characteristics. Similar remarks apply to hens, their eggs, and chicks which are kept in buildings without sunlight or with ordinary glass windows; these, also, do not survive, and so pass out of the picture.

With humans, the situation is entirely different. We are able to produce conditions which do not exist in unaided nature. We produce artificial conditions and so *our numbers and distribution* are not regulated by unaided nature alone. Animals cannot over-populate the globe, as they do not produce artificially. We do over-populate this globe because we produce artificially. With *animals, selfishness comes before altruism, and the non-selfish perish. An animal has to live first, then act.* With man, the reverse is true. The selfish may produce such conditions that they are destroyed by them. We can over-populate the globe because of artificial production, and so we are actually born nowadays into a world where we must *act first before we can live.* As I have already shown in my *Manhood of Humanity* (p. 72), the old animalistic, fallacious generalizations have been, and are, the foundation of our 'philosophies', 'ethics', systems., and naturally such animalistic doctrines must be disastrous to us. Neurologically, we build up conditions which our nervous systems cannot stand; and so we break down, and, perhaps, shall not even survive.

Animals have no 'doctrines' in *our meaning* of the term; thus, doctrines are no part of their environment, and, accordingly, animals cannot perish through false doctrines. We do have them, however, and, since they are the most vital environmental semantic conditions regulating our lives, if they are fallacious, they make our lives unadjusted and so, ultimately, lead to non-survival.

So we see that 'human adjustment' is quite a different and much more complex affair than 'animal adjustment'. The 'world' of 'man' is also a different and much more complex 'world' than that of the animal. There seems to be no escape from this conclusion. We see, also, that what we used to call 'senses' supply us with information about the world that is very limited in quantity, *specific* in quality, an abstraction of low order, never being 'it'. Being often unaffected, our 'senses' are not able to abstract, obviously, some of the most fundamental manifestations of energy to be found in the external world. If we speak of the older so-called 'sense perceptions' as lower order abstractions, then we find that we learn about the other subtler manifestations of energy through science, higher neural and extra-neural means, which we may call higher order abstractions. In the older days, we called this kind of knowledge 'inferential knowledge'. The animals do not have these higher order abstractions in that sense, and so their world is for them devoid of these extra-sensible manifestations of energy.

It must be remembered that these higher order abstractions and the 'inferential knowledge' of the old theories (they are not equivalent by definition) have a very similar status. Organisms work as-a-whole, and to separate completely higher and lower order abstractions is impossible. All that is said here justifies the new terminology. Our nervous system does abstract, does summarize, does integrate on different levels and in different orders, and the *result* of a stimulus *is not* the stimulus itself. The stone *is not* the pain produced by the stone dropping on our foot; neither is the flame we see, nor the burn we feel. The actual process goes on outside of our skin, as represented by the 'realities' of modern science.

We have already spoken frequently of the different order abstractions, their special characteristics, dynamic versus static., and the means of translation of lower orders into higher, and vice versa. Events which are going on and for which we have no direct 'senses' of abstraction, as, for instance, electric waves, Röntgen-rays, wireless waves., we know only through extra-neural extensions of our nervous system given by science and scientific instruments. Naturally, we should expect that the structure of our abstracting mechanism would be also reflected in these higher order abstractions. Facts show this to be true; and practically all modern science proves it directly or indirectly. This is why, for instance, we have the mathematical methods for passing from dynamic to static, and vice versa; why we have quantum theories and conditions; and why we have problems of continuity versus discontinuity, atomic theories, .

The above is not a plea for certain old-fashioned 'idealistic philosophies' and still less for 'solipsism'. Far from it. The object of this present work is to face hard structural experimental *m.o* facts, analyse these facts in a language of a similar *structure* ([non-A]), and so to reach tentatively new conclusions which again can be verified by experiments. Once more the reader must be warned against carelessly translating the structurally *new* terms into the *old* terms. The complete structural, psycho-logical, semantic, and neurological analysis of one single such new term would afford material for several volumes and so is impossible in this work. The usefulness of the old terms has been exhausted. The structural consequences of the old terms have been practically all worked out, and, as a rule, we cannot have much quarrel with the older conclusions in the *old language*. If we reach *different* conclusions, or get some new emphasis, it will be due-to the use of the structurally new language. If we translate the new into the old, the old conclusions are usually *truer* than the new ones. The reverse is also true; the old conclusions become false or, at best, only gain emphasis because of the structure of the new language. The problem of all theories, old or new, is to give a structural account of the facts known, to account for exceptions, and to predict new experimental structural facts which again may be verified empirically.

9. MATHEMATICS AS A LANGUAGE OF A STRUCTURE SIMILAR TO THE STRUCTURE OF THE WORLD

The present work—namely, the building of a *non-aristotelian system,* and an introduction to a theory of sanity and general semantics—depends, fundamentally, for its success on the recognition of mathematics as a language similar in structure to the world in which we live.

It is a common experience of our race that with a happy generalization many unconnected parts of our knowledge become connected; many 'mysteries' of science become simply a linguistic issue, and then the mysteries vanish. New generalizations introduce new *attitudes* (evaluation) which, as usual, seriously simplify the problems for a new generation. In the present work, we are treating problems from the point of view of such a generalization, of wide application; namely, *structure,* which is forced upon us by the denial of the 'is' of identity.; so that structure becomes the only link between the objective and verbal levels. The next consequence is that structure alone is the only possible content of knowledge.

Investigating structure, we have found that structure can be defined in terms of relations; and the latter, for special purposes, in terms of multi-dimensional order. Obviously, to investigate structure, we must look for relations, and so for multi-dimensional order. The full application of the above principles becomes our guide for future enquiry.

A structural independent analysis of mathematics, treated as a language and a form of human behaviour, establishes the similarity of this language to the undeniable structural characteristics of this world and of the human nervous system. These few and simple structural foundations are arrived at by inspection of known data and may be considered as well established.

From the point of view of general semantics, mathematics, having symbols and propositions, must be considered as a language. From the psychophysiological point of view, it must be treated as an activity of the human nervous system and as a form of the behaviour of the organisms called humans.

All languages are composed of two kinds of words: (1) Of *names* for the somethings on the un-speakable level, be they external objects., or *internal feelings,* which admittedly are *not* words, and (2) of *relational terms,* which express the actual, or desired, or any other relations between the un-speakable entities of the objective level.

When a 'quality' is treated physiologically as a reaction of an organism to a stimulus, it also becomes a relation. It should be noticed that often some words can be, and actually are, used in both senses; but, in a given context, we can always, by further analysis, separate the words used into these two categories. Numbers are not exceptions; we can use the labels 'one', 'two'., as numbers (of which the character will be explained presently) but also as names for anything we want, as, for instance, Second or Third Avenue, or John Smith I or John Smith II. When we use numbers as names, or labels for anything, we call them numerals; and this is *not* a mathematical use of 'one', 'two'., as these names do not follow mathematical rules. Thus, Second Avenue and Third Avenue cannot be added together, and do not give us Fifth Avenue in any sense whatever.

Names alone do not produce propositions and so, by themselves, say nothing. Before we can have a proposition and, therefore, meanings, the names must be related by some relation-word, which, however, may be explicit or implied by the context, the situation, by established habits of speech, . The division of words into the above two classes may seem arbitrary, or to introduce an unnecessary complication through its simplicity; yet, if we take modern knowledge into account, we cannot follow the grammatical divisions of a primitive-made language, and such a division as I have suggested above seems structurally correct in 1933.

Traditionally, mathematics was divided into two branches: one was called arithmetic, dealing with numbers; the other was called geometry, and dealt with such entities as 'line', 'surface', 'volume', . Once Descartes, lying in bed ill, watched the branches of a tree swaying under the influence of a breeze. It occurred to him that the varying distances of the branches from the horizontal and vertical window frames could be expressed by numbers representing measurements of the distances. An epoch-making step was taken: geometrical relations were expressed by numerical relations; it meant the beginning of analytical geometry and the unification and arithmetization of mathematics.

Further investigation by the pioneers Frege, Peano, Whitehead, Russell, Keyser, and others has revealed that 'number' can be expressed in 'logical terms'—a quite important discovery, provided we have a valid 'logic' and structurally correct *non-el* terms.

Traditionally, too, since Aristotle, and, in the opinion of the majority, even today, mathematics is considered as uniquely connected with quantity and measurement. Such a view is only partial, because there are many most important and fundamental branches of mathematics which have nothing to do with quantity or measurement—as, for instance, the theory of groups, analysis situs, projective geometry, the theory of numbers, the algebra of 'logic', .

Sometimes mathematics is spoken about as the science of relations, but obviously such a definition is too broad. If the only content of knowledge is structural, then relations, obvious, or to be discovered, are the foundation of all knowledge and of all language, as stated in the division of words given above. Such a definition as suggested would make mathematics co-extensive with all language, and this, obviously, is not the case.

A semantic definition of mathematics may run somehow as follows: Mathematics consists of limited linguistic schemes of multiordinal relations capable of exact treatment at a given date.

After I have given a semantic definition of number, it will be obvious that the above definition covers all existing disciplines considered mathematical. However, these developments are not fixed affairs. Does that definition provide for their future growth? By inserting as a fundamental part of the definition 'exact treatment at a given date', it obviously does. Whenever we discover any relations in any fields which will allow exact 'logical' treatment, such a discipline will be included in the body of linguistic schemes called mathematics, and, at present, there are no indications that these developments can ever come to an end. When 'logic' becomes an ∞-valued 'structural calculus', then mathematics and 'logic' will merge completely and become a general science of *m.o* relations and multi-dimensional order, and all sciences may become exact.

It is necessary to show that this definition is not too broad, and that it eliminates notions which are admittedly non-mathematical, without invalidating the statement that the content of all knowledge is structural, and so ultimately relational. The word 'exact' eliminates non-mathematical relations. If we enquire into the meaning of the word 'exact', we find from experience that this meaning is not constant, but that it varies with the date, and so only a statement 'exact at a given date' can have a definite meaning.

We can analyse a simple statement, 'grass is green' (the 'is' here is the 'is' of predication, not of identity) , which, perhaps, represents an extreme example of a non-mathematical statement; but a similar reasoning can be applied to other examples. Sometimes we have a feeling which we express by saying, 'grass is green'. Usually, such a feeling is called a 'perception'. But is such a *process* to be dismissed so simply, by just calling it a name, 'perception'? It is easy to 'call names under provocation', as Santayana says somewhere; but does that exhaust the question?

If we analyse such a statement further, we find that it involves comparison, evaluation in certain respects with other characters of experience., and the statement thus assumes relational characteristics. These, in the meantime, are non-exact and, therefore, non-mathematical. If we carry this analysis still further, involving data taken from chemistry, physics, physiology, neurology., we involve relations which become more and more exact, and, finally, in such terms as 'wave-length', 'frequency'., we reach structural terms which allow of exact 1933 treatment. It is true that a language of 'quality' conceals relations, sometimes very effectively; but once 'quality' is taken as the reaction of a given organism to a stimulus, the term used for that 'quality' becomes a name for a very complex relation. This procedure can be always employed, thus establishing once more the fundamental character of relations.

These last statements are of serious structural and semantic importance, being closely connected with the [non-A], fundamental, and undeniable negative premises. These results can be taught to children very simply; yet this automatically involves an entirely new and modern method of evaluation and attitude toward language, which will affect beneficially the, as yet entirely disregarded, *s.r.*

We must consider, briefly, the terms 'kind' and 'degree', as we shall need them later. Words, symbols., serve as forms of representation and belong to a different universe—the 'universe of discourse'—since they are not the un-speakable levels we are speaking about. They belong to a world of higher abstractions and not to the world of lower abstractions given to us by the lower nerve centres.

Common experience and scientific investigations (more refined experience) show us that the world around us is made up of absolute individuals, each different and unique, although interconnected. Under such conditions it is obviously optional what language we use. The more we use the language of diverse 'kind', the sharper our definitions must be. Psycho-logically, the emphasis is on difference. Such procedure may be a tax on our ingenuity, but by it we are closer to the structural facts of life, where, in the limit, we should have to establish a 'kind' for every individual.

In using the term 'degree', we may be more vague. We proceed by similarities, but such a treatment implies a fundamental interconnection between different individuals of a special kind. It implies a definite kind of metaphysics or structural assumptions—as, for instance, a theory of evolution. As our 'knowledge' is the result of nervous abstracting, it seems, in accordance with the structure of our nervous system, to give preference to the term 'degree' first, and only when we have attained a certain order of verbal sharpness to pass to a language of 'kind', if need arises.

The study of primitive languages shows that, historically, we had a tendency for the 'kind' language, resulting in over-abundance of names and few relation-words, which makes higher analysis impossible. Science, on the other hand, has a preference for the 'degree' language, which, ultimately, leads to mathematical languages, enormous simplicity and economy of words, and so to better efficiency, more intelligence, and to the unification of science. Thus, chemistry became a branch of physics, physics, a branch of geometry; geometry merges with analysis, and analysis merges with general semantics; and life itself becomes a physico-chemical colloidal occurrence. The language of 'degree' has very important *relational, quantitative,* and *order* implications, while that of 'kind' has, in the main, qualitative implications, often, if not always, concealing relations, instead of expressing them.

The current definition of 'number', as formulated by Frege and Russell, reads: 'The number of a class is the class of all those classes that are similar to it'.[1] This definition is not entirely satisfactory: first, because the multiordinality of the term 'class' is not stated; second, it is [A], as it involves the ambiguous (as to the order of abstractions) term 'class'. What do we mean by the term 'class'? Do we mean an extensive array of absolute individuals, un-speakable by its very character, such as some *seen aggregate*, or do we mean the *spoken definition* or *description* of such un-speakable objective entities? The term implies, then, a fundamental confusion of orders of abstractions, to start with—the very issue which we must avoid most carefully, as positively demanded by the non-identity principle. Besides, if we explore the world with a 'class of classes'., and obtain results also of 'class of classes', such procedure throws no light on mathematics, their applications and their importance as a tool of research. Perhaps, it even increases the mysteries surrounding mathematics and conceals the relations between mathematics and human knowledge in general.

We should expect of a satisfactory definition of 'number' that it would make the semantic character of numbers clear. Somehow, through long experience, we have learned that numbers and measurement have some mysterious, sometimes an uncanny, importance. This is exemplified by mathematical predictions, which are verified later empirically. Let me recall only the discovery of the planet Neptune through mathematical investigations, based on its action upon Uranus, long before the astronomer actually verified this prediction with his telescope. Many, a great many, such examples could be given, scientific literature being full of them. Why should mathematics and measurement be so extremely important? Why should mathematical operations of a given Smith, which often seem innocent (and sometimes silly enough) give such an unusual security and such undeniably practical results?

Is it true that the majority of us are born mathematical imbeciles? Why is there this general fear of, and dislike for, mathematics? Is mathematics really so difficult and repelling, or is it the way mathematics is treated and taught by mathematicians that is at fault? If some light can be thrown on these perplexing semantic problems, perhaps we shall face a scientific revolution which might deeply affect our educational system and may even mark the beginning of a new period in standards of evaluation, in which mathematics will take the place which it ought to have. Certainly, there must be something the matter with our epistemologies and 'psychologies' if they cannot cope with these problems.

A simple explanation is given by a new [non-A] analysis and a *semantic* definition of numbers. What follows is written, in the main, for non-mathematicians, as the word 'semantic' indicates, but it is hoped that professional mathematicians (or some, at least) may be interested in the *meanings* of the term 'number', and that they will not entirely disregard it. As semantic, the definition seems satisfactory; but, perhaps, it is not entirely satisfactory for technical purposes, and the definition would have to be slightly re-worded to satisfy the technical needs of the mathematicians. In the meantime, the gains are so important that we should not begrudge any amount of labour in order to produce finally a mathematical and, this time, [non-A] semantic definition of numbers.

As has already been mentioned, the importance of notation is paramount. Thus the Roman notation for number—I, II, III, IV, V, VI ., — was not satisfactory and could not have led to modern developments in mathematics, because it did not possess enough positional and structural characteristics. Modern mathematics began when it was made possible by the invention or discovery of positional notation. We use the symbol '1' in 1, 10, 100, 1000., in which, because of its place, it had different values. In the expression '1', the symbol means 'one unit'; in 10, the symbol '1' means ten units; in 100, the symbol '1' means one hundred units, .

In the evolution of mathematics, we find that the notions of 'greater', 'equal', and 'less' precede the notion of numbers. Comparison is the simplest form of evaluation; the first being a search for relations; the second, a discovery of exact relations. This process of search for relations and structure is inherent and natural in man, and has led not only to the discovery of numbers, but also has shaped their two aspects; namely, the cardinal and the ordinal aspects. For instance, to ascertain whether the number of persons in a hall is equal to, greater than, or less than the number of seats, it is enough to ascertain if all seats are occupied and there are no empty seats and no persons standing; then we would say that the number of persons is equal to the number of seats, and a *symmetrical relation* of equality would be established. If all seats were occupied and there were some persons standing in the hall, or if we found that no one was standing, yet not all seats were occupied, we would establish the *asymmetrical relation* of greater or less.

In the above processes, we were using an important principle; namely, that of *one-to-one* correspondence. In our search for relations, we assigned to each seat one person, and reached our conclusions without any counting. This process, based on the *one-to-one* correspondence, establishes what is called the cardinal number. It gives us specific relational data about this world; yet it is not enough for counting and for mathematics. To produce the latter, we must, first of all, establish a definite system of symbolism, based on a definite relation for generating numbers; for instance, $1+1=2$, $2+1=3$., which establishes a definite *order*. Without this ordinal notion, neither counting nor mathematics would be possible; and, as we have already seen, order can be used for defining relations, as the notions of relation and order are interdependent. Order, also, involves asymmetrical relations.

If we consider the two most important numbers, 0 and 1, we find that in the accepted symbolism, if $a=b$, $a-b=0$; and if $a=b$, $a/b=b/b=1$; so that both fundamental numbers express, or can be interpreted as expressing, a *symmetrical relation* of equality.

If we consider any other number—and this applies to all kinds of numbers, not only to natural numbers—we find that any number is not altered by dividing it by one, thus, $2/1=2$, $3/1=3$., in general, $N/1=N$; establishing the *asymmetrical* relation, *unique* and *specific* in a given case that N is N times more than one.

If we consider, further, that $2/1=2$, $3/1=3$, *and so on*, are all *different*, *specific*, and *unique*, we come to an obvious and [non-A] semantic definition of number in terms of relations, in which 0 and 1 represent *unique* and *specific symmetrical* relations and all other numbers also *unique* and *specific asymmetrical* relations. Thus, if we have a result '5', we *can always say* that the number 5 is five times as many as one. Similarly, if we introduce apples. Five apples are five times as many as one apple. Thus, a number in any form, 'pure' or 'applied', can always be represented as a relation, *unique* and *specific* in a given case; and this is the foundation of the exactness of dealing with numbers. For instance, to say that *a is* greater than *b* also establishes an asymmetrical relation, but it is *not unique* and *specific*; but when we say that *a* is five times greater than b, this relation is *asymmetrical, exact, unique,* and *specific.*

It seems that mathematicians, no matter how important the work which they have produced, have never gone so far as to appreciate fully that they are willy-nilly producing an ideal human relational language of structure similar to that of the world *and* to that of the human nervous system. This they cannot help, in spite of some vehement denials, and their work should also be treated from the semantic point of view.

Similarly with measurement. From a functional or actional and semantic point of view, measurement represents nothing else but a search for *empirical structure* by means of extensional, ordered, symmetrical, and asymmetrical relations. Thus, when we say that a given length measures five feet, we have reached this conclusion by *selecting* a unit called 'foot', an *arbitrary and un-*

speakable affair, then laying it end to end five times in a definite *extensional order* and so have established the asymmetrical, and, in each case, *unique* and *specific* relation, that the given entity represents, in this case, five times as many as the arbitrarily selected unit.

As the only possible content of knowledge is structural, as given in terms of relations and multiordinal and multi-dimensional order; numbers, which establish an endless array of exact, specific, and, in each case, unique, relations are obviously the most important tools for exploring the *structure* of the world, since structure can always be analysed in terms of *relations*. In this way, all mysteries about the importance of mathematics and measurement vanish. The above understanding will give the student of mathematics an entirely different and a very natural feeling for his subject. As his only possible aim is the study of structure of the world, or of whatever else, he must naturally use a *relational tool* to explore this complex of relations called 'structure'. A most spectacular illustration of this is given in the internal theory of surfaces, the tensor calculus., described in Part VIII.

In all measurements, we select a unit of a necessary kind, for a given case, and then we find a *unique* and *specific* relation as expressed by a number, between the given something and the selected unit. By relating different happenings and processes to the same unit-process, we find, again, *unique* and *specific* interrelations, in a given case, between these events, and so gather structural (and most important, because uniquely possible) wisdom, called 'knowledge', 'science', .

If we treat numbers as relations, then fractions and all operations become relations of relations, and so relations of higher order, into the analysis of which we cannot enter here, as these are, of necessity, technical.

Let me again emphasize that, from time immemorial, things have *not* been words; the only content of knowledge has been structural; mathematics has dealt, in the main, with numbers; no matter whether we have understood the character of numbers or not, numbers have expressed relations and so have given us structural data willy-nilly, . This explains why mathematics and numbers have, since time immemorial, been a favorite field, not only for speculations, but also why, in history, we find so many religious semantic disturbances connected with numbers. Mankind has somehow felt instinctively that in numbers we have a potentially endless array of *unique* and *specific exact relations*, which ultimately give us structure, the last being the only possible content of knowledge, because words are not things.

As relations, generally, are empirically present, and as man and his 'knowledge' is as 'natural' as rocks, flowers, and donkeys, we should not be surprised to find that the unique language of exact., relations called mathematics is, by necessity, the natural language of man and *similar in structure* to the world *and* our nervous system.

As has already been stated, it is incorrect to argue from the structure of mathematical theories to the structure of the world, and so try to establish the similarity of structure; but that such enquiry must be *independent* and start with quite ordinary structural experiences, and only at a later stage proceed to more advanced knowledge as given by science. Because this analysis must be independent, it can also be made very simple and elementary. All exact sciences give us a wealth of experimental data to establish the first thesis on similarity of structure; and it is unnecessary to repeat it here. I will restrict myself only to a minimum of quite obvious facts, reserving the second thesis—about the similarity of structure with our nervous system—for the next chapter.

If we analyse the silent objective level by objective means available in 1933, say a microscope, we shall find that whatever we can see, handle., represents an *absolute individual*, and *different from anything else in this world*. We discover, thus, an important *structural* fact of the external world; namely, that in it, everything we can see, touch., that is to say, all lower order abstractions represent absolute individuals, different from everything else.

On the verbal level, under such empirical conditions, we should then have a language of *similar structure*; namely, one giving us an indefinite number of *proper names, each different*. We find such a language *uniquely* in numbers, each number 1, 2, 3., being a unique, sharply distinguishable, *proper name* for a relation, and, if we wish, for anything else also.

Without some higher abstractions we cannot be human at all. No science could exist with absolute individuals and no relations; so we pass to higher abstractions and build a language of say x_i, ($i = 1, 2, 3, \ldots n$), where the x shows, let us say, that we deal with a variable x with many values, and the number we assign to i indicates the individuality under consideration. From the structural point of view, such a vocabulary is similar to the world around us; it accounts for the individuality of the external objects, it also is similar to the structure of our nervous system, because it allows generalizations or higher order abstractions, emphasizes the abstracting nervous characteristics,. The subscript emphasizes the differences; the letter x implies the similarities.

In daily language a similar device is extremely useful and has very far-reaching psychological semantic effects. Thus, if we say 'pencil$_1$', 'pencil$_2$', . . . 'pencil$_n$', we have indicated structurally two main characteristics: (1) the absolute individuality of the object, by adding the indefinitely individualizing subscript 1, 2, . . . n; and (2) we have also complied with the nervous higher order abstracting characteristics, which establish similarity in diversity of different 'pencils'. From the point of view of *relations*, these are usually found empirically; besides, they may be invariant, no matter how changing the world may be.

In general terms, the structure of the external world is such that we deal always on the objective levels with absolute individuals, with absolute differences. The structure of the human nervous system is such that it abstracts, or generalizes, or integrates., in higher orders, and so finds similarities, discovering often invariant (sometimes relatively invariant) relations. To have 'similar structure', a language should comply with both structural exigencies, and this characteristic is found in the mathematical notation of x_i, which can be enlarged to the daily language as 'Smith$_i$', 'Fido$_i$', where $i = 1, 2, 3, \ldots n$.

Further objective enquiry shows that the world and ourselves are made up of *processes*, thus, 'Smith$_{1900}$' is quite a different person from 'Smith$_{1933}$'. To be convinced, it is enough to look over old photographs of ourselves, the above remark being structurally entirely general. A language of 'similar structure' should cover these facts. We find such a language in the vocabulary of 'function', 'propositional function', as already explained, involving also four-dimensional considerations.

As words are not objects—and this expresses a structural fact—we see that the 'is' of identity is unconditionally false, and should be entirely abolished as such. Let us be simple about it. This last semantic requirement is genuinely difficult to carry out, because the general *el* structure of our language is such as to facilitate identification. It is admitted that in some fields some persons identify only a little; but even they usually identify a great deal when they pass to other fields. Even science is not free from identification, and this fact introduces great and artificial semantic difficulties, which simply vanish when we stop identification or the confusion of orders of abstractions. Thus, for instance, the semantic difficulties in the foundations of mathematics, the problems of 'infinity', the 'irrational'., the difficulties of Einstein's theory, the difficulties of the newer quantum theory, the arguments about the 'radius of the universe', 'infinite velocities', the difficulties in the present theory., ., are due, in the main, to semantic blockages or commitment to the structure of the old language—we may call it 'habit'—*which says structurally very little*, and which I disclose as a semantic disturbance of *evaluation* by showing the *physiological mechanism in terms of order.*

If we abolish the 'is' of identity, then we are left only with a functional, actional., language elaborated in the mathematical language of function. Under such conditions, a *descriptive* language of ordered happenings on the objective level takes the form of 'if so and so happens, then so and so happens', or, briefly, 'if so, then so'; which is the prototype of 'logical' and mathematical processes and languages. We see that such a language is again similar in structure to the external world descriptively; yet it is similar to the 'logical' nervous processes, and so allows us, because of this similarity of structure, predictability and so rationality.

In the traditional systems, we did not recognize the complete semantic interdependence of differences and similarities, the empirical world exhibiting differences, the nervous system manufacturing primarily similarities, and our 'knowledge', if worth anything at all, being the *joint product* of both. Was it not Sylvester who said that 'in mathematics we look for similarities in differences and differences in similarities'? This statement applies to our whole abstracting process.

The empirical world is such in structure (by inspection) that in it we can add, subtract, multiply, and divide. In mathematics, we find a language of similar structure. Obviously, in the physical world these actions or operations alter the relations, which are expressed as altered unique and specific relations, by the language of mathematics. Further, as the world is full of different shapes, forms, curves., we do not only find in mathematics special languages dealing with these subjects, but we find in analytical geometry unifying linguistic means for translation of one language into another. Thus any 'quality' can be formulated in terms of relations which may take the 'quantitative' character which, at present, in all cases, can be also translated into geometrical terms and methods, giving structures to be *visualized.*

It is interesting, yet not entirely unexpected, that the activities of the higher nervous centres, the conditional reflexes of higher order, the semantic reactions, time-binding included, should follow the exponential rules, as shown in my *Manhood of Humanity.*

In our experience, we find that some issues are additive—as, for instance, if one guest is added to a dinner party, we will have to add plates and a chair. Such facts are covered by additive methods and the language called 'linear' (see Part VIII). In many instances—and these are, perhaps, the most important and are strictly connected with submicroscopic processes—the issues are not additive, one atom of oxygen 'plus' two atoms of hydrogen, under proper conditions, will produce water, of which the characteristics are not the sum of the characteristics of oxygen and hydrogen 'added' together, but entirely *new* characteristics *emerge*. These may some day be taken care of by non-linear equations, when our knowledge has advanced considerably. These problems are unusually important and vital, because with our present low development and the lack of structural researches, we still keep an additive [A] language, which is, perhaps, able to deal with additive, simple, immediate, and comparatively unimportant issues, but is entirely unfit structurally to deal with principles which underlie the most fundamental problems of life. Similarly, in physics, only since Einstein have we begun to see that the primitive, simplest, and easiest to solve linear equations are not structurally adequate.

One of the most marked structural characteristics of the empirical world is 'change', 'motion', 'waves', and similar dynamic manifestations. Obviously, a language of similar structure must have means to deal with such relations. In this respect, mathematics is unique, because, in the differential and integral calculus, the four-dimensional geometries and similar disciplines, with all their developments, we find such a perfect language.

10. MATHEMATICS AS A LANGUAGE OF A STRUCTURE SIMILAR TO THE STRUCTURE OF THE HUMAN NERVOUS SYSTEM

General

Mathematics in the twentieth century is characterized by an enormous productiveness, by the revision of its foundations, and the quest for rigour, all of which implies material of great and unexplored psychological value, a result of the activity of the human nervous system. Branches of mathematics, as, for instance, mathematical 'logic', or the analytical theory of numbers, have been created in this period; others, like the theory of function, have been revised and reshaped. The theory of Einstein and the newer quantum mechanics have also suggested further needs and developments.

Any branch of mathematics consists of propositional functions which state certain structural relations. The mathematician tries to discover new characteristics and to reduce the known characteristics to a dependence on the smallest possible set of constantly revised and simplest structural assumptions. Of late, we have found that no assumption is ever 'self-evident' or ultimate.

To those structural assumptions, we give at present the more polite name of postulates. These involve undefined terms, not always stated explicitly, but always present implicitly. A *postulate system gives us the structure of the linguistic scheme.* The older mathematicians were less particular in their methods. Their primitive propositional functions or postulates were less well investigated. They did not start explicitly with undefined terms. The twentieth century has witnessed in this field a marked progress in mathematics, though much less in other verbal enterprises; which accounts for the long neglect of the structure of languages. Without tracing down a linguistic scheme to a postulate system, it is extremely difficult or impossible to find its structural assumptions.

A peculiarity of modern mathematics is the insistence upon the formal character of all mathematical reasoning, which, with the new *non-el* theory of meanings, ultimately should apply to all linguistic procedures.

The problems of 'formalism' are of serious and neglected psychological importance, and are connected with great semantic dangers in daily life if associated with the lack of consciousness of abstracting; or, in other words, when we confuse the orders of abstractions. Indeed, the majority of 'mentally' ill are *too formal* in their psycho-logical, one-, two-, or few-valued processes and so cannot adjust themselves to the ∞-valued experiences of life. Formalism is only useful in the search for, and test of, structure; but, in that case, the consciousness of abstracting makes the attitude behind formal reasoning ∞-*valued and probable*, so that semantic disturbances and shocks in life are avoided. Let us be simple about it: the mechanism of the semantic disturbance, called 'identification', or 'the confusion of orders of abstractions' in general, and 'objectification' in particular, is, to a large extent, dependent on two-valued formalism without the consciousness of abstracting.

In mathematics, formalism is uniquely useful and necessary. In mathematics, the formal point of view is pressed so far as to disclaim that any meanings, in the ordinary sense, have been ascribed to the undefined terms, the emphasis being on the postulated relations between the undefined terms. The last makes the majority of mathematicians able to adjust themselves, and mathematics extremely general, as it allows use to ascribe to the mathematical postulates an indefinite number of meanings which satisfy the postulates.

This fact is not a defect of mathematics; quite the opposite. It is the basis of its tremendous practical value. It makes mathematics a linguistic scheme which embodies the possibility of perfection, and which, no doubt, satisfies semantically, at each epoch, the great majority of properly informed individual Smiths and Browns. There is nothing absolute about it, as all mathematics is ultimately a product of the human nervous system, the best product produced at each stage of our development. The fact that mathematics establishes such linguistic relational patterns without specific content, accounts for the generality of mathematics in applications.

If mathematics had physical content or a definite meaning ascribed to its undefined terms, such mathematics could be applied only in the given case and not otherwise. If, instead of making the mathematical statement that one and one make two, without mentioning what the one or the two stands for, we should establish that one apple and one apple make two apples, this statement would not be applied safely to anything else but apples. The generality would be lost, the validity of the statement endangered, and we should be deprived of the greatest value of mathematics. Such a statement concerning apples is not a mathematical statement, but belongs to what is called 'applied mathematics', which has content. Such experimental facts as that one gallon of water added to one gallon of alcohol gives *less* than two gallons of the mixture, do not invalidate the mathematical statement that one and one make two, which remains valid by definition. The last mentioned experiment with the 'addition' of water to alcohol is a deep sub-microscopic structural characteristic of the empirical world, which must be discovered at present by experiment. The most we can say is that we find the above mathematical statement applicable in some instances, and non-applicable in others.

Not assigning definite meanings to the undefined terms, mathematical postulates have variable meanings and so consist of propositional functions. Mathematics must be viewed as a manifold of patterns of exact relational languages, representing, at each stage, samples of the best working of the human 'mind'. The application to practical problems depends on the ingenuity of those desiring to use such languages.

Because of these characteristics, mathematics, when studied as a form of human behaviour, gives us a wealth of psycho-logical and semantic data, usually entirely neglected.

As postulates consist of propositional functions with undefined terms, all mathematical proof is formal and depends exclusively on the form of the premises and not on special meanings which we may assign to our undefined terms. This applies to all 'proof'. 'Theories' represent linguistic structures, and must be proved on semantic grounds and never by empirical 'facts'. Experimental facts only make a theory more plausible, but no number of experiments can 'prove' a theory. A proof belongs to the *verbal* level, the experimental facts do not; they belong to a different order of abstractions, not to be reached by language, the connecting link being *structure*, which, in languages, is given by the systems of postulates.

Theories or doctrines are always linguistic. They formulate something which is going on inside our skin in relation to what is going on on the un-speakable levels, and which is not a theory. Theories are the rational means for a rational being to be as rational as he possibly can. As a fact of experience, the working of the human nervous system is such that we have theories. Such was the survival trend; and we must not only reconcile ourselves with this fact, but must also investigate the structure of theories.

Theories are the result of extremely complex cyclic chains of nerve currents of the human nervous system. Any semantic disturbance, be it a confusion of orders of abstractions, or identification, or any of their progeny, called 'elementalism', 'absolutism', 'dogmatism', 'finalism'., introduces some deviations or resistances, or semantic blockages of the normal survival cycles, and the organism is at once on the abnormal non-adjustment path.

Mathematics as a Language of a Structure Similar to the Structure of the Human Nervous System

The structure of protoplasm of the simplest kind, or of the most elaborate nervous system, is such that it abstracts and reacts in its own specific way to different external and internal stimuli.

Our 'experience' is based normally on abstractions and integrations of different stimuli by different receptors, with different and specific reactions. The eye produces its share, and we may see a stone; but the eye does not convey to us the *feel* of weight of the stone, or its temperature, or its hardness, . To get this new wisdom, we need other receptors of an entirely different kind from those the eye can supply. If the eye plays some role in establishing the weight, for instance, without ever giving the *actual feel* of weight, it is usually misleading. If we would try to lift a pound of lead and a pound of feathers, which the balance would register as of equal weight, the pound of lead would feel heavier to us than the pound of feathers. The eye saw that the pound of lead is smaller in bulk, and so the doctrinal, semantic, and muscular expectation was for a smaller weight, and so, by contrast, the pound of lead would appear unexpectedly heavy.

As the eye is one of the most subtle organs, in fact, a part of the brain, science is devising methods to bring all other characteristics of the external world to direct or indirect inspection of the eye. We build balances, thermometers, microscopes, telescopes, and other instruments, but the character and *feel* of weight, or warmth., must be supplied directly by the special receptors, which uniquely can produce the special 'sensations'. The swinging of the balance, or the rise of the column of the thermometer, establishes most important *relations*, but does not give the immediate specific and un-speakable feel of 'weight' or of 'warmth'. Our first and most primitive contact with a stone, its feel., is a personal abstraction from the object, full of characteristics supplied by the peculiarities of the special receptors. Our primitive picture 'stone' is a summary, an integration, of all these separate 'sense' abstractions. It is an abstraction from many abstractions, or an abstraction of a higher order.

Theories are relational or structural verbal schemes, built by a process of high abstractions from many lower abstractions, which are produced not only by ourselves but by others (time-binding). Theories, therefore, represent the shortest, simplest structural summaries and generalizations, or the highest abstractions from individual experience and through symbolism of racial past experiences. Theories are mostly not an individual, but a collective, product. They follow a more subtle but inevitable semantic survival trend, like all life. Human races and epochs which have not revised or advanced their theories have either perished, or are perishing.

The process of abstracting in different orders being inherent in the human nervous system, it can neither be stopped nor abolished; but it can be deviated, vitiated, and forced into harmful channels contrary to the survival trend, particularly in connection with pathological s.r. No one of us, even when profoundly 'mentally' ill, is free from theories. The only selection we can make is between antiquated, often primitive-made, theories, and modern theories, which always involve important semantic factors.

The understanding of the above is of serious importance, as, by proper selection of theories, all wasteful semantic disturbances, which lead even to crimes, and such historical examples of human un-sanity as the 'holy inquisition', burning at the stake, religious wars, persecution of science, the Tennessee trial., could have been avoided.

Whenever *any one* says *anything*, he is indulging in theories. A similar statement is true of writing or 'thinking'. We *must* use terms, and the very selection of our terms and the structure of the language selected reflect their structure on the subject under discussion. Besides, words are not the events. Even simple 'descriptions', since they involve terms, and ultimately undefined terms, involve structural assumptions, postulates, and theories, conscious or unconscious—at present, mostly the latter.

It is very harmful to sanity to teach a disregard for theories or doctrines and theoretical work, as we can never get away from them as long as we are humans. If we disregard them, we only build for ourselves semantic disturbances. The difference between morbid and not so obviously morbid confusions of orders of abstractions is not very clear. The strong affective components of such semantic disturbances must lead to absolutism, dogmatism, finalism, and similar states, which are semantic factors out of which states of un-sanity are built.

We know that we must start with undefined terms, which may be defined at some other date in other undefined terms. At a given date, our undefined terms must be treated as postulates. If we prefer, we may call them structural assumptions or hypotheses. From a theoretical point of view, these undefined terms represent not only postulates but also variables, and so generate propositional functions and not propositions. In mathematics, these issues are clear and simple. Every theory is ultimately based on postulates which consist of propositional functions containing variables, and which express relations, indicating the structure of the scheme.

It appears that the main importance of the linguistic higher order abstractions is in their *public* character, for they are capable of being transmitted in neural and extra-neural forms. But our private lives are influenced also very much by the lower order abstractions, 'feelings', 'intuitions',. These can be, should be, but seldom are, properly influenced by the higher order abstractions. These 'feelings'., are personal, unspeakable, and so are non-transmittable. For instance, we cannot transmit the actual feeling of pain when we burn ourselves; but we can transmit the invariant relation of the extremely complex fire-flesh-nerve-pain manifold. A relation is present empirically, but also can be expressed by words. It seems important to have means to translate these higher order abstractions into lower.

The Psychological Importance of the Theory of Aggregates and the Theory of Groups

Starting with the [non-A] denial of identity, we were compelled to consider structure as the only possible link between the empirical and the verbal worlds. The analysis of structure involved relations and *m.o* and multi-dimensional order, and, ultimately, has led us to a semantic definition of mathematics and numbers. These definitions make it obvious that all mathematics expresses general processes of mentation *par excellence*. We could thus review all mathematics from this psycho-logical point of view, but this would not be profitable for our purpose; so we will limit ourselves to a brief sketch connected with the theory of aggregates and the theory of groups, because these two fundamental and most general theories formulate in a crisp form the general psycho-logical process, and also show the mechanism by which all languages (not only mathematics) have been built. Besides, with the exception of a few specialists, the general public is not even aware of the existence of such disciplines which depart widely from traditional notions about mathematics. They represent most successful and powerful attempts at building exact relational languages in subjects which are on the border-line between psycho-logics and the traditional mathematics. Because they are exact, they have been embodied in mathematics, although they belong just as well to a general science of relations, or general semantics, or 'psychology', or 'logic', or scientific linguistics and psychophysiology. There are other mathematical disciplines, as, for instance, analysis situs, or the 'algebra of logic'., to which the above statements apply; but, for our present purposes, we shall limit ourselves to the former two.

Dealing with the theory of aggregates, I will give only a few definitions taken from the *Encyclopaedia Britannica*, with the purpose of drawing the attention of the 'psychologists', and others, to those psychological data.

Mathematics as a Language of a Structure Similar to the Structure of the Human Nervous System

The theory of aggregates underlies the theory of function. An aggregate, or manifold, or set, is a system such that: (1) It includes all entities to which a certain characteristic belongs; and (2) no entity without this characteristic belongs to the system; (3) any entity of the system is permanently recognizable as distinct from other entities.

The separate entities which belong to such a collection, system, aggregate, manifold, or set are called elements. We assume the possibility of selecting at pleasure, by a definite process or law, one or more elements of any aggregate A, which would form another aggregate B, .

The above few lines express how the human 'thought' processes work and how languages were built up. It is true that the exactness imposes limitations, and so the mathematical theories are not expressed in the usual antiquated 'psychological' terms, although they describe one of the most important psycho-logical processes.

Lately, the theory of aggregates has led to a weighty question: Does one of the fundamental laws of old 'logic'; namely, the two-valued law of the 'excluded third' (A is either B or not B), apply in all instances? Or is it valid in some instances and invalid in others?

This problem is the psycho-logical kernel of the new revision of the foundation of mathematics, which has lately been considerably advanced by Professor Lukasiewicz and Tarski with their many-valued 'logic', which merges ultimately with the mathematical theory of probability; and on different grounds has perhaps been solved in the present *non-el*, [non-A]-system.

The notion of a group is psycho-logically still more important. It is connected with the notions of transformation and invariance. Without giving formal definitions unnecessary for our purpose, we may say that if we consider a set of elements a, b, c., and we have a rule for combining them, say O, and if the result of combining any two members of the set is itself a member of the set, such aggregate is said to have the 'group property'.

Thus, if we take numbers or colours, for instance, and the rule which we accept is '+', we say that a number or a colour is transformed by this rule into a number or a colour, and so both possess the 'group property'. Obviously, by performing the given operation, we have transformed one element into another; yet some characteristics of our elements have remained invariant under transformation. Thus, if 1 is a number and 2 is a number, the operation '+' transforms 1 into 3, since 1 + 2 = 3; but 3 has the character of being a number; so this characteristic is preserved or remains invariant. Similarly, with colours, if we add colours, these are transformed, but remain colours, and so both sets have the 'group property'. Keyser suggests that the 'mental' processes have the group property, which is undoubtedly true.

The role this theory plays in our language is of great importance, because in it we find a method of search for structure, and a method by which we can establish a similarity of structure between the un-speakable objective level and the verbal level, based on invariance of relations which are found or discovered in both.

In the notion of a group, we have become acquainted with two terms; namely, transformation and invariance. The first implies 'change'; the other, a lack of 'change' or 'permanence'. Both of these characteristics are semantically fundamental, but involve serious complexities.

The world, ourselves included, can be considered as processes which can be analysed in terms of transformed stages with all their derivative notions. In the objective world, 'change' is ever present and is, perhaps, the most important structural characteristic of our experience. But when a highly developed nervous system, a process itself, is acted upon by other processes, such nervous system discovers, at some stage of its development, a certain relative permanence, which, at a still later stage, is formulated as invariance of function and relations. The latter formulation is *non-el* because it can be discovered empirically, which means by the lower nerve centres, but also is the main necessity and means of operating of the higher nerve centres, so-called 'thought',. All that we usually call a process of 'association' is nothing else than a *process of relating*, a direct consequence of the structure of the nervous system, where stimuli are registered in a certain four-dimensional order, which, on the psycho-logical level, take the form of relations. From this point of view, it is natural that the higher nerve centres, as a limit of integrating processes, should produce *and* discover invariance of relations, which appears then as the supreme product and so, ultimately, a necessity of the activity of the higher centres. Obviously, if the invariance of relations has any objective counterpart whatsoever in the external world, this invariance is impressed on the nervous system more than other characteristics; and so, at a certain stage, a nervous system which is capable of producing and using a highly developed symbolism, must discover and formulate this invariance.

It seems that *relations*, because of the possibility of discovering them and their invariance in *both* worlds, are, in a way, more 'objective' than so-called objects. We may have a science of 'invariance of relations', but we could not have a science of permanence of things; and the older doctrines of the permanence of our institutions must also be revised. Under modern conditions, which change rather rapidly nowadays, obviously, some relations between humans alter, and so the institutions must be revised. If we want *their invariance*, we must build them on such *invariant relations between humans* as are not altered by the transformations. This present work, indeed, is concerned with investigating such relations, and they are found in the *mechanism of time-binding*, which, once stated, becomes quite obvious after reflection.

As Professor Shaw says: 'We find in the invariants of mathematics a source of objective truth. So far as the creations of the mathematician fit the objects of nature, just so far must the inherent invariants point to objective reality. Indeed, much of the value of mathematics in its applications lies in the fact that its invariants have an objective meaning. When a geometric invariant vanishes, it points to a very definite character in the corresponding class of figures. When a physical invariant vanishes or has particular values, there must correspond to it physical facts. When a set of equations that represent physical phenomena have a set of invariants or covariants which they admit, then the physical phenomena have a corresponding character, and the physicist is forced to explain the law resulting. The unnoticed invariants of the electromagnetic equations have overturned physical theories, and have threatened philosophy. Consequently the importance of invariants cannot be too much magnified, from a practical point of view'.

It should be noticed that the *non-el* character of the terms relation, invariance., which apply both to 'senses' and 'mind', is particularly important, as it allows us to apply them to all processes; and that such a language is similar in structure not only to the world around us, but also to our nervous processes. Thus, a process of being iron, or a rock, or a table, or you, or me, may be considered, for practical purposes, as a temporal and average invariance of function on the sub-microscopic level. Under the action of other processes, the process becomes structurally transformed into different relational complexes, and we die, and a table or rock turns into dust, and so the invariance of this function vanishes.

The notion of a function involves the notion of a variable. The functional notion has been extended to the propositional function and, finally, to the doctrinal function and system-function. The term transformation is closely related to that of function and relation. This notion is based on our capacity to associate, or relate, any two or more 'mental' entities. We can, for instance, associate *a* with *b* or *b* with *a*. We say that we have transformed *a* into *b*, or vice versa.

An excellent example of transformation, given by Keyser, is an ordinary dictionary, which would be genuinely mathematical if it were more precise. In a dictionary, every word is transformed into its verbal meaning, and vice versa. A telephone directory is another example. Quite obviously, the term 'transformation' has far-reaching implications. If *a* is transformed into *b*, this implies that there is a relation between *a* and *b* which is being established, by the fact of transformation. Once a relation is established, we have a propositional function of two or more variables which define an extensional set of all elements connected by this relation.

We see that these three terms are inseparably united and are three aspects of one psycho-logical process. If we have a transformation, we have a function and a relation; if we have a function, we have a relation and a transformation; if we have a relation, we have a transformation and a function. Transformation, as we see, is a psycho-logical term of action. A relation has a psycho-logically mixed character. A propositional function is a static statement, on record, with blanks for the values of the variables. In it the form is invariant, but it may take an indefinite number of values. The *extensional manifold of* the values for the variable is static, given once for all in a given context. It is extensional and, therefore, may be empirical and experimental.

Let us take as an example, for instance, the transformation of a set of integers 1, 2, 3, . Let us suppose that the given law of transformation is given by the function $y=2x$. The result would be the manifold of even integers 2, 4, 6, . We see that integers are transformed into integers; therefore, the characteristic of being an integer is preserved; in other words, this characteristic is an invariant under the given transformation $y=2x$, but the values of the integers are not preserved.

The theory of invariance is an important branch of mathematics, made famous of late through the work of Einstein. Einstein fulfilled the dearest dream of Riemann and attained the methodological and scientific ideal, that a 'law of nature' should be formulated in such a manner as to be invariant under groups of transformations. Such a semantic ideal, once stated, cannot be denied; it expresses exactly a necessity of the proper working of the human nervous system. In fact, a 'law' of nature represents nothing else than a statement of the invariance of some relations. When the Einstein criterion is applied, it renders most of the old 'natural laws' invalid, as they cannot stand the test of invariance. The older 'universal laws' then appear as local private gossips, true for one observer and false for another.

The method of the theory of invariance gives us the trend of relations that abide, and so expresses important psycho-logical characteristics of the human 'mind'. Its further significance is revealed by Keyser in the suggestion that when a group of transformations leaves some specified psycho-logical activity invariant, it defines perfectly some actual or potential branch of science, some actual or potential doctrine.

We all know how deeply rooted in us is the feeling, the longing for stability, how worried we are when things become unstable. Worries and fear are destructive to semantic health and should be taken into account in a theory of sanity. A similar semantic urge apparently moved mathematicians when they worked out the theory of invariance; it was a formulation of a necessity of the activities of the human nervous system. That similar semantic methods, if applied, would give similar results in our daily lives, scarcely needs to be emphasized.

We have already spoken of the mathematical theory of invariance as a mathematical species of a semantic theory of universal agreement. Similarly, in a [non-A]-system based on relations and structure, it is possible to formulate a theory of universal agreement which would be structurally impossible in the [A]-system, and so the dreams of Leibnitz become a sober reality; but we must first re-educate our *s.r.*

Similarity in Structure of Mathematics and the Human Nervous System

In the chapter on the Semantics of the Differential Calculus, the fundamental notions and method of this calculus are explained. Here we may say, briefly, that it consists in stratifying, or expanding into a series, of an interval of any sort which proceeded by large steps. The large steps are divided into a great number of smaller and smaller steps, which, in the limit, when the numbers of steps become infinite, take on the aspect of 'continuity' so that we can study the 'rate of change'. When 'time' is taken into consideration, the dynamic may be translated into static, and vice versa; processes can be analysed at any stage,. This short description is far from exact or exhaustive; I emphasize only in an intuitive way what is of main semantic importance for our purpose.

The main object of the present chapter is to explain that the structure of the human nervous system is such that, on some levels, we produce dynamic abstractions; on others, static. As the organism works as-a-whole, for its optimum working, and, therefore, for sanity, we need a language, a method, which may be translated into a *s.r* by which to translate the dynamic into the static, and vice versa; and such a language, such a method, is produced and supplied by mathematicians. To some readers, these remarks may appear so obvious as to make it unnecessary to write them, but I have found, through personal observation of reactions of different individuals, and by a careful survey of the literature of the subject, that even many mathematicians and physicists do not have this *s.r* in all problems—or, at least, they do not know how to apply it.

In Part VII, elementary [non-A] methods are worked out, which supply the neurological semantic benefits of the calculus, very easily imparted to even small children *without any mathematical technique*, and establishing in them a mathematical attitude toward all language in general, training them in the only structural psycho-logics of sanity; namely, that of the calculus, which thus becomes the foundation of healthy and normal human *s.r*. And this, let us repeat again, without any mathematical technique. We find, also, that there are simple and *physiological* means, based on structure, of training our *s.r* and imparting the feel for the structural stratification inherent in the consciousness of abstracting.

To start with, let me mention briefly a quite unexpected, unconscious, structural *biological* characteristic of mathematics; namely, its (in the main) *non-el*, organism-as-a-whole character.

From the time of Aristotle, biologists, physiologists, neurologists, 'psychologists', psychiatrists and others have spoken a great deal about the organism-as-a-whole; yet, they have not seemed to realize that if they produce *el* terms, they cannot apply the *non-el* principle.

It will probably not be an exaggeration to say that the majority of mathematicians have never heard of this principle, and that, if they have, they paid no attention to it; *yet*, in practice, they have applied it very thoroughly. The main mathematical terms are *non-el*, organism-as-a-whole terms which apply to 'senses' as well as to 'mind'. For instance, relation, order, difference, variable, function, transformation, invariance., can mostly be seen as well as 'thought' of. The use of such terms prevents our speculation from degenerating into purely *el* speculations on words, a process always closely related to the morbid semantic manifestations of the 'mentally' ill, and obviously based on the pathological confusion of orders of abstractions, involving inappropriate evaluation.

This fact alone is of serious importance, as it indicates that mathematics is a language of similar structure to the structure of organisms and is a correct language, not only neurologically, but also *biologically*. This characteristic of mathematics, quite unexpectedly discovered, made the fusion of geometry and physics possible. It underlies, also, the theory of space-time and the

Einstein theory. It will be seen later that it has also serious psycho-neurological importance.

It was already emphasized that the existing 'psychologies' are animalistic or metaphysical, because either they disregard one of the most unique human characteristics, such as the behaviour called mathematizing, or they indulge in speculations on, and in, *el* terms. It was suggested that no *human* 'psychologist' can actually perform his official task unless he is an equipped student of mathematics. Unless we actually apply the *non-el* principle, and take into account that the structure of languages introduces implications, unconscious in the main, and that no man is ever free from some doctrines and some so-called 'logical' processes involving physiological and semantic concomitants, no general theory of *human* 'psychology' can be produced.

The above solves a very knotty semantic problem, for we see that if we apply the *non-el* principle, any 'psychology' on the human level must become *psycho-logics*, though the old term 'psychology' could be retained as applying to animal researches only. The very name 'psychology', or the 'theory or science of mind', is obviously *el*, and treats 'mind' as an objective separate entity. As these results were originally reached independently, it is interesting to notice that the modern methods and the application of the structural positive knowledge 1933 lead to very many analogies and similarities, though this, after all, might be expected.

Notice the hyphen which, out of the *el* and delusional objectified 'space' and 'time', made the einsteinian space-time a language of *non-el* structure similar to the world around us; and the hyphen which out of *el* 'psychology' makes a *non-et* human discipline of psycho-logics. It seems that a little dash here and there may be of serious semantic importance when we deal with symbolism.

To facilitate exposition, it is useful to stress, in the present section, the neurological and psychiatrical side, as an outline of the methods of the calculus, and related subjects will, of necessity, require separate treatment.

When rats are trained to perform a simple experiment requiring some 'mentality' and afterwards a large part of the cerebral cortex is removed, their training may be wholly lost. If such decorticated rats are trained again, they re-acquire the habit as readily as before. It appears that, with rats, the cortex is not essential for these learning processes. They 'learn' as well, or nearly as well, with their sub-cortical and thalamic regions. In what follows, to avoid misstatements, I will use the rather vague term, yet sufficient for my purpose, 'thalamic region' or 'lower centres' instead of more specific terms, the use of which would complicate the exposition unnecessarily. With dogs, apes, and men, the situation is increasingly different. Their nervous systems are more differentiated. Their functional interchangeability is impaired. In the most complex human brain there still exists some interchangeability of function. When an arm, for instance, is paralysed through a brain-lesion, the arm may re-acquire a nearly normal function, though there is no regeneration of the destroyed brain tissue. However, the interchangeability is less pronounced than in the lower brains. There seems to be ho doubt that the thalamic regions are not only a vestibule through which all impulses from the receptors have to pass in order to reach the cortex, but also that the affective characteristics are strictly connected with processes in these regions. It seems that some very primitive and simple associations can be carried on by the thalamic regions.

The cortex receives its material as elaborated by the thalamus. The abstractions of the cortex are abstractions from abstractions and so ought to be called abstractions of higher order. In neurology, similarly, the neurons first excited are called of 'first order'; and the succeeding members of the series are called neurons of the 'second order',. Such terminology is structurally similar to the inherent structure and function of the nervous system. The receptors are in direct contact with the out-side world and convey their excitation and nerve currents to the lower nerve centres, where these impulses are further elaborated and then abstracted by the higher centres.

According to our daily experience and scientific knowledge, the outside world is an ever-changing chain of events, a kind of flux; and, naturally, those nerve centres in closest contact with the outside world must react in a shifting way. These reactions are easily moved one way or another, as in our 'emotions', 'affective moods', 'attention', 'concentration', 'evaluation', and other such semantic responses. In these processes, some associative or relational circuits exist, and there may be some very low kind of 'thinking' on this level. Birds have a well-developed, or, perhaps, over-developed, thalamus but under-developed and poor cortex, which may be connected with their stupidity and excitability.

Something similar could be said about the 'thalamic thinking' in humans; those individuals who overwork their thalamus and use their cortex too little are 'emotional' and stupid. This statement is not exaggerated, because there are experimental data to show how through a psycho-neural training the *s.r*, in some cases, can be re-educated, and that with the elimination of the semantic disturbances there is a marked development of poise, balance, and a proportional increase of critical judgement, and so 'intelligence'. Idiots, imbeciles, and morons are usually 'emotional' and excitable, as well as deficient in their 'mental' processes. A similar characteristic can be found in other unclassified 'mentally' deficient, and their name is legion—a characteristic strictly connected with, and often produced by, disturbances of the *s.r.* When these shifting, dynamic, affective, thalamic-region, lower order abstractions are abstracted again by the higher centres, these new abstractions are further removed from the outside world and must be somehow different.

In fact, they *are* different; and one of the most characteristic differences is that they have *lost* their *shifting* character. These new abstractions are relatively static. It is true that one may be supplanted by another, but they do not change. In this fact lies the tremendous value and danger of this mechanism, as disclosed clearly by the disturbances of the *s.r.* The value is chiefly in the fact that such higher order abstractions represent a perfected kind of memory, which can be recalled exactly in the form as it was originally produced. For instance, the circle, *defined* as the locus of points in a plane at equal distance from a given point called the centre, remains permanent as long as we wish to use this definition. We can, therefore, recall it perfectly, analyse it., without losing the definiteness and the stability of this memory. Thus, critical analysis, and, therefore, progress, becomes possible. Compare this perfected memory, which may last indefinitely unchanged, with memories of 'emotions' which, whether dim or clear, are always distorted. We see that the first are reliable, that the others are not.

Another most important characteristic of the higher order abstractions is that, although of neural origin, they may be preserved and used over and over again in extra-neural forms, as recorded in books and otherwise. This fact is never fully appreciated from a neurological point of view. Neural products are stored up or preserved in extra-neural form, and they can be put back in the nervous system *as active neural processes*. The above represents a fundamental mechanism of time-binding which becomes overwhelmingly important, provided we discover the physiological mechanism of regulating the *s.r*, on the one hand, and discover the mechanism by which these extra-neural factors can be made physiologically effective, on the other.

If humans are characterized by the fact that they build up this cumulative affair called 'civilization', this is possible through those higher order abstractions and the time-binding ability to extend our nervous system by extra-neural means, which, in the meantime, may play a most important neural role and become active nervous impulses. The last is only possible if some abstractions are static, and so can be recorded, leading ultimately to further extensions of the human nervous system by extra-neural means, such as microscopes, telescopes, and practically all modern scientific instruments, books, and other records.

To illustrate what has been said here, I know of no better example than is found in moving pictures. When we watch a moving picture representing some life occurrence, our 'emotions' are aroused, we 'live through', the drama; but the details, in the main, are blurred, and a short time after seeing it either we forget it all or in parts, or our memory falsifies most effectively what was seen. It is easy to verify the above experimentally by seeing one picture twice or three times, with an interval of a few days between each seeing. The picture was 'moving', all was changing, shifting, dynamic, similar to the world *and* our feelings on the un-speakable levels. The impressions were vague, shifting, non-lasting, and what was left of it was mostly coloured by the individual mood., while seeing the moving picture. Naturally, under such conditions, there is little possibility of a rational scientific analysis of a situation.

But if we *stop* the moving film which ran, say, thirty minutes, and analyse the static and extensional series of small pictures on the reel, we find that the drama which so stirred our 'emotions' in its moving aspect becomes a series of slightly different static pictures, each difference between the given jerk or grimace being a *measurable* entity, establishing relations which last indefinitely.

The *moving* picture represents the usually brief processes going on in the lower nerve centres, 'close to life', but unreliable and evading scrutiny. The *arrested* static film which lasts indefinitely, giving *measurable* differences between the recorded jerks and grimaces, obviously allows analysis and gives a good analogy of the working of higher nerve centres, disclosing also that all life occurrences have many aspects, the selection of which is mostly a problem of our pleasure and of the selection of language. The moving picture gives us the process; each static film of the reel gives us stages of the process in chosen intervals. In case we want a moving picture of a growing plant, for instance, we photograph it at given intervals and then run it in a moving-picture projector, and then we see the process of growth. These are empirical facts, and the calculus supplies us with a language of similar structure with many other important consequences.

We know that a number of human races have perished without leaving many traces of their existence. This process is going on continually, even now. Some races are progressing; some are regressing; some are at a seeming standstill. It would appear that the mechanism of higher order abstractions had and has survival value, and, therefore, should not be neglected but cultivated. In this special case, cultivation is a condition inherent in the process and a necessity for time-binders.

Serious semantic dangers are also revealed by analysis and verified by observation. These higher order abstractions, let us repeat, are static and may last indefinitely, as long as for structural reasons we do not replace the old by new ones. Even then, though rejected, they remain as a permanent fact on record. Obviously, these higher abstractions have only a 'second-hand' connection with the outside world. Even their character is changed, they are static while the world is dynamic. The lower 'sense' world has 'characteristics left out', owing to the mechanism of abstracting of the lower centres; and the abstractions of higher orders have 'all characteristics included', because these are abstractions from abstractions, an *intra-organismal* process in its entirety, their starting material being already an end-product of the activities of the lower centres. This mechanism is only under full control if we are conscious of abstracting, because the higher order abstractions in the nervous chain affect, in their turn, the lower centres, and, in pathological cases, impress on them a semantic *delusional* or *illusional* evaluation as if a character of experience. In severe cases, even the lower nerve centres are stimulated to such an extent that hallucinations appear.

If we do not know how to handle different order abstractions, this results in serious semantic dangers. If the distribution of the returning nerve currents is a non-survival one, we exhibit semantic disturbances, such as identification or confusion of orders of abstractions, delusions, illusions, and hallucinations. Thus, we ascribe to the products of the lower nerve centres, the lower order abstractions, characteristics fictitious and impossible for them, such as 'immutability', 'permanence', involving disorientation about 'time'., ., which are characteristics of the higher order abstractions, but do not belong to the world as given by the lower abstractions, and result in an improper evaluation disturbing to the *s.r.* Such disturbances make us, naturally, absolutists and dogmatists, involve serious affective disturbances, and lead to non-adaptive behaviour and reactions, and other semantic manifestations of un-sanity. These, in their turn, make adjustment more difficult, often affecting the structure of man-made institutions, which again make adjustments more complex and often impossible. We become un-sane, 'insane', and life, whether public or private, becomes a mess. In such a vicious semantic circle, we distort our education, our systems, and institutions. Often the morbid reactions of powerful individuals are forced upon masses, who are then ruled by these morbid products, with injury to their nervous systems. Different mass hysterias, 'revivals', wars, political and religious propaganda, very often commercial advertisements, offer notable examples.

The morbid semantic influence of commercialism has not been investigated, but it does not take much imagination to see that commercial psycho-logics, as exemplified by the theories of commercial evaluation, 'wisdom', appeal to selfishness, animal cunning, concealing of true facts, appeal to 'sense' gratification., produce a *verbal and semantic environment* and slogans for the children which, if preserved in the grown-ups, must produce some pathological results. It is hoped that some day a psychiatrist will investigate this large, neglected, and very important semantic problem.

The lack of structural linguistic researches and investigation of our *s.r*, and the ignorance of those who rule, make us nearly helpless. Malaria or other germ diseases would never be eliminated were we to preserve religiously the sources of infection. The semantic sources of un-sanity are not only defended but are actively sponsored by organized ignorance and the power of merchants, state, and church.

The situation is acute. If we could entirely eliminate our cortex, it would, perhaps, not be so serious. We could, perhaps, live as complex a life as a fish and have a nervous system perfectly adjusted to such a life. But, unfortunately, with a structural change, or, according to Lashley, with the change even in the total mass of the brain, the activities and the role of the whole, including other parts, are profoundly altered.[8] These become inadequate, as shown by the boy born without the cortex, already described. His nervous system was much more complex than that of fishes or of some lower animals which lead *adequately* a rather complex life. But the boy was less equipped for life than they. Even his 'senses', though apparently 'normal' on macroscopic levels, must have been pathological on colloidal and sub-microscopic levels and did not function properly. We know, also, that in many cases of 'mental' ills the 'sense reactions' are abnormal; sometimes the patients seem to be entirely insensitive to stimuli which would produce most acute pain to other less pathological individuals.

It is impossible to eliminate completely from our lives or nerve currents the higher abstractions and their psycho-neural effect. Curiously enough, this elementary fact has never been emphasized or taken into account seriously; yet it is a crucial semantic factor in our attitude toward science and our future. Those who attempt such elimination, whether by actively persecuting science, or by emitting propaganda against science, or by the cynical or ignoring attitude toward 'mental' achievements, whether personally, or in education, or in public prints, or other public activities, do not succeed in eliminating the higher order abstractions, but simply introduce *pathological semantic reactions* and succeed in disorganizing their own nervous systems and those of others. I intended this implication when I said that our existing educational., systems *produce* morons, but 'geniuses' are born. Such very general semantic directives are, perhaps, responsible for the extremely low level of our non-technical development. Humans are not to be judged simply by the ability to drive an automobile or by the knowledge of how to use a bathtub; nor yet by their capacity for buying and selling things produced by others.

The tendency of some public prints to appeal to the morbidity of mob psycho-logics and to its ignorance, insisting that all that is said should be said in 'one-syllable' words, so that the mob can understand, in a human class of life, is an *arresting* or *regressive tendency.* What should be urged for sanity, and for humans, is that the mob should also learn the use of at least two-syllable words! Then, perhaps, the day would come when they could follow easily and habitually the use of *non-el* terms and, perhaps, even of words connected by a hyphen.

This appeal to mob psycho-logics and ignorance affects profoundly our *s.r* and should be investigated. It definitely appears that in countries where the majority reads only the sort of publications referred to above and commercial advertisements, their psycho-logical equipment and standards are lower than those of perfectly illiterate peasants of other countries. It is not fully realized that in a symbolic class of life, symbolism of any sort—e.g., public prints—plays an environmental role and creates *s.r* which may be distinctly morbid. The problems of public prints, commercialism., and their psycho-logical effect on the *s.r* should undergo a searching analysis by psychiatrists, and definite suggestions should be formulated by psychiatric scientific organizations or congresses.

Under the conditions prevailing at present, it is futile to preach 'morals' of any metaphysical kind. They have never worked satisfactorily, and increasingly they cannot work, particularly under the present much more complex conditions of life. They disorganize the survival activities and processes of the human nervous system. The imposed and delusional dogmas are themselves the result of pathological evaluation in their originators; a necessity, perhaps, on a primitive level, but profoundly semantically harmful under the complexities of life-conditions 1933.

As it is impossible to eliminate the influence of the higher order abstractions, we should investigate whether or not we can control these processes and the related *s.r.* We can learn to regulate these processes, which otherwise may become pathological, and to redirect the currents into constructive survival channels. I can state definitely that this is possible. We can control physiologically the *s.r* through the elimination of identification, by training in order, in consciousness of abstracting, and similar disciplines, and thus eliminate the pathological semantic disturbances of confusion of orders of abstractions. Such training, whenever possible, has seemingly a beneficial influence even on the more extreme pathological states listed above, and suggests general preventive value.

Let me briefly restate the fundamental differences between lower order abstractions and higher. The lower order abstractions are manufactured by the lower nerve centres, which are closer to, and in direct contact with, actual life experiences. These are non-permanent, shifting, vague and un-speakable, but often very intense. They play a most important role in our daily lives. They cannot be transmitted, as they are essentially of a non-transmittable character, and have a private, non-public character. All 'sense' impressions, 'feelings', 'moods'., are representative. of them. We should remember that, detached, they are fictions, manufactured verbally, because our language happens to be *el*. Actually, these lower centres are in the cyclic chain and so influence, and are influenced by, the full cycle, including the higher order abstractions, whatever the latter may be in a given individual. The main point is that they are shifting, changing, non-permanent, non-stable— 'moving', so to say—and remain un-speakable.

The higher order abstractions are abstractions from the lower order abstractions, being further removed from the outside world, and are of a distinctly different character. These are static, 'permanent', and cannot be entirely eliminated from any one.

From the point of view of sanity, the problem of how we can handle these functions becomes paramount. In the cyclic nervous chain, we always must translate one level into the other. Obviously, if, in the *higher* centres, we elaborate shifting, changing, non-permanent material, this material is not appropriate for them; they cannot work properly, and some pathological processes may set in.

If we elaborate the *lower* nerve centres abstractions that are static, permanent., in character, and hence inappropriate for the lower centres, we build up morbid non-survival identifications, delusions, illusions, hallucinations, and other disturbances of evaluation, resulting in milder cases in absolutism, dogmatism, fanaticism., and, in heavier cases, in a neurosis or even a psychosis.

It seems quite obvious that each nervous level has its own specific kind of material to deal with. As they are in a cyclic nervous chain and are interconnected in a bewilderingly complex way, the problem of appropriate translation of one level of abstractions into the other becomes a semantic foundation for a well-balanced functioning of the nervous system. In this respect, we differ fundamentally from animals. The above difficulties do not arise in animals to that extent, because their nervous systems are not differentiated enough for such sharp differentiation in the functioning. For this reason, without human interference, there could be no 'insane' animals which could survive (see Part VI). But, having no static higher order abstractions in the human sense, they cannot pass on their 'experiences', which are transmittable *only* in the higher order formulations in neural and extra-neural forms to the next generations. Animals are not time-binders.

For humans, the proper translation of dynamic into static and static into dynamic becomes paramount for sanity, on psycho-logical levels, affecting, probably by colloidal processes, the psycho-neural foundation of semantic responses.

Psychiatry informs us that most of the 'mentally' ill have their main disturbances in the dynamic affective field. It is a very difficult field to reach by the older methods, the more so that the older *el* sharp distinction between 'intellect' and 'emotions' prevented the discovery of workable means. 'Thinking' and 'feeling' are not to be divided so simply. We know how 'thinking' is influenced by 'feeling'; but we know very little how 'feeling' is influenced by 'thinking'—perhaps, because we have not analysed the semantic issues in *non-el* terms.

All psychotherapy, with its manifold theories, each contributing its share, is a semantic attempt to influence 'feeling' by 'thinking'. A large number of successful cases seems to show clearly that some such means are possible. Large numbers of failures show equally that the methods used are not structurally satisfactory. The need of more scientific investigations of a more general and fundamental, *non-el* character becomes emphatic. The present enquiry shows that such structural investigations suggest that the method can be found in the psycho-logics of the 'mind' at its best; namely, in mathematics, which unexpectedly leads to a physiological control of the *s.r*, effective not only as a therapeutic, but also as a preventive, educational means.

Identification as a factor of un-sanity seems to be a natural consequence of the evolution from 'animal' to 'man', particularly at our present stage, while the human race is so recent a product. The human cortex appeared only comparatively lately and is a young structure; the thalamic regions have a much longer history of functioning. It seems natural that the nervous impulses should pass the shorter, more phylogenetically travelled, paths in preference to comparatively newer and longer paths, a principle well known in neurology in connection with so-called 'Bahnung'. If education, and on human levels any kind of adjustment involving *s.r* involves some education, fails to force the nerve currents into their proper channels, or actively establishes in them semantic psycho-neural blockages through pathological evaluation acquired because of faulty training, we should expect either infantilism or regression to still lower levels. Whatever the correct explanation of the distribution of nerve currents, semantic blockages., may be, observation shows unmistakably that some such assumptions are necessitated by observed manifestations in behaviour. Experiments show, also, that such defects can be helped greatly by the proper re-training and re-education of the *s.r*.

To understand the structure of these semantic disturbances, we must become acquainted with the affective components which underlie mathematics and mathematical methods, hitherto disregarded, because of the *el* character of our old terminology. There is another striking connection. In severe 'mental' illnesses, we usually find a disorientation in 'space' and 'time', which are, by necessity, *relational data* of experience. In the semantic disturbances called identification, we also find, as a rule, relational disorientation *about* 'space' and 'time', more subtle but very vicious in effect, bordering on what are called 'philosophical' problems, which, as a matter of fact, represent psycho-neural disturbances. Since Einstein, the disturbances can be easily eliminated, provided we take into account structural *non-el* issues in connection with *s.r* and a [non-A]-system.

Up to this point, we have been emphasizing the beneficial structural aspect of mathematics, and it is now necessary to explain why mathematizing, when considered as a formal interplay of contentless symbols should not be considered a high-class 'mental' activity, no matter how useful and important it may be, and why the majority of mathematicians do not get the *full* psycho-logical semantic benefit of their training and activities. The nervous systems of many such mathematicians do not act fully and successfully, nor pass normally through the cycle of their natural activities. Such a technician is seldom, if ever, what we call a great man. He seldom has a direct creative influence on our lives. But, in the case of a man with a more efficient nervous system, the cycle is completed successfully, the higher abstractions are translated back into new lower abstractions, which are closer to life. Such an individual 'sees', 'visualizes', has 'intuitions'., in his symbolic interplays. He then has a new structural vision through a new survey of his own experiences and all the experiences of others when translated in terms of lower centres. He gains a deeper insight, which he ultimately makes useful to all of us.

Immediate experience, always un-speakable, is strictly connected with the lower centres. In the translation of experience into higher order abstractions and language, the un-speakable character of experience is lost, and a *new neurological process* is needed to re-translate these higher order abstractions into new lower abstractions, and thus fully and successfully to complete the nervous cycle. One can learn to play with symbols according to rules, but such play has little creative value. If the translation is made into the language of lower centres—namely, into 'intuitions', 'feelings', 'visualizations'., —the higher abstractions gain the character of experience, and so creative activity begins. Individuals with thoroughly efficient nervous systems become what we call 'geniuses'. They create new values by inventions of new methods and in other ways, which give us a new structural means of exploring, and thus of dealing with, the world around us and ourselves, and so, ultimately, human adjustment is helped.

It is important for the reader to become thoroughly familiar with the simple division of our nervous processes into terms of order in a cyclic chain. Even neurology calls the neurons excited first of 'first order', and the succeeding members of the series, of 'second order', . The above considerations have an important practical semantic bearing for all of us, since many of the processes which we are describing can be influenced educationally by simple methods, because the term 'order', when applied, acquires a *physiological* character for *evaluation*. The description and verbal analysis of the process is, naturally, complex, but once the physiological base of evaluation is discovered, the training becomes very simple, although not easy.

The principal aim of this present work is to make available a simple and practical physiological means for accomplishing what is highly desirable, and, at the same time, for eliminating what is semantically undesirable. We deal with mathematics, because mathematics is *unique*, and, being unique, has no substitute. When discussing the theory of meanings, we have shown that all verbalism is, ultimately, similar to mathematics in structure. This conclusion contradicts many current theories of language and meanings, and so, at this stage of our argument, we lay special emphasis on the only discipline in which these issues are clear and obvious; namely, mathematics. The older theories, based on ignorance of mathematics, have led to serious abuses of our linguistic capacities and to s.r which are mostly pathological, with the result that practically 99 per cent of us are semantically disturbed and un-sane. Many of us, even, are on the verge of more serious 'mental' illnesses.

It will be well to give a rough picture of the similarities of, and differences between, the working of the human 'mind' at its worst ('insanity'), and its working at its best (mathematics). We shall find that the average man is between the two, often dangerously close to the first. The following picture is rough and one-sided, but suggestive, and should be worked out more fully.

The 'insane' have structural, conscious or unconscious, 'premises', which are 'false', or, in general, semantically inappropriate. Their s.r are shifting when they should be static, or static when they should be flexible. In the main, the difficulty of evaluation lies in the lower abstractions and the affective field. These abstractions are not properly transmitted or translated or regulated by the higher centres; or else, the higher order static abstractions are projected with too strong affective components on the lower centres. Hence, different identifications, delusions, illusions, and hallucinations result. Their 'ideas' are evaluated as things or experience, and affectively objectified in different degrees, which results in the above mis-evaluating manifestations. These semantic disturbances and tensions make the 'mentally' ill believe irresistibly in the 'truth' of their 'premises' and their inductions and deductions, which they follow blindly. In them, as in the rest of us, some internal affective pressure comes first, but because in humans the effect of higher nerve centres cannot be entirely abolished, this affective pressure is rationalized somehow into some sort of 'premises'. This organism-as-a-whole process is entirely general and applies to all of us in all our activities, but is most clearly seen in the ordered details in the work of creative scientists and 'geniuses', and in the more severe cases of 'mental' illness. To the 'mentally' ill these 'premises' have the value of 'the' and not 'a' premise. *They act upon them,* and so cannot adjust themselves to a world different from their fancies. They would seldom survive at all if left alone by themselves, particularly in a complex 'civilization'.

Mathematicians, also, have structural premises, often called postulates, but they *never* evaluate them to be 'true'; wherefore their premises *cannot* be 'false'. They have no claims, and claims are always affective. Like the 'insane', they follow up these premises blindly, but, being generally conscious of abstracting in the field of their profession, they are not usually subject to semantic disturbances *in this field* and do not live out their theories in life, the theories thus remaining affectively hypothetical. If a mathematician were to believe, with strong affective evaluation, that his premises are 'true', these premises then would become mostly false, or, meaningless, or, in general, inappropriate. If he lived through them, the given individual would then be 'mentally' ill, not because of his premises, but because of the semantic disturbance, which would involve erroneous evaluation, identifications, confusion of orders of abstractions in his affective *attitude toward his premises*. This subtle organism-as-a-whole mechanism, in which all affective pressure can be rationalized, and all rationalization can produce affective manifestations, not only makes the present *non-el* analysis possible and legitimate, but also offers some explanation of those remarkable cases of 'mental' illness in a number of mathematical geniuses. Under such organism-as-a-whole structural conditions, a *general* consciousness of abstracting not restricted to a special field is the only possible safeguard against the semantic disturbances which lead to an unbalanced 'mental' condition.

As we have seen, the difference between 'sanity' and 'insanity' is subtle. The reader must be reminded that it takes a good 'mind' to be 'insane'. Morons, imbeciles, and idiots are 'mentally' deficient, but could not be 'insane'.

The so-called 'sane' also have structural premises; we all have some standards of evaluation. These are also usually false, or, in general, inappropriate, being mostly due to our savage inheritance. But the saner we are, the less we abide by them. Therefore, in a world quite different structurally from our fancies, we are often able to adjust ourselves for all practical purposes, often avoiding major disasters for a number of years.

For instance, the believers in extraordinary blisses in the 'other life' or the 'other world' should welcome death. Why be so unhappy here, when, according to their doctrines, there is such an ideally happy future after death? Why make use of medicine and doctors, when a deadly illness should open the door to everlasting bliss! In conflict with such a creed, he lives as long as he can, often most unhappily, and is generally willing to spend fortunes on doctors and medicines to delay the bliss! The genuine and very serious danger to all of us of such creeds is that when the *s.r* of an individual are trained in this way he finally does become indifferent, or apathetic toward actualities in *this world*, so that cunning, and often pathological, individuals are thus given an opportunity of directing human affairs toward their personal ends.

Naturally, with the increase of the complexities of conditions, the dangers also increase in a geometrical ratio, because when *m.o* realities become too unbearable, the masses cease to be influenced by these semantic illusions, and they break all barriers, only to fall again under the influence of new leaders very often equally irresponsible and ignorant.

Unfortunately, the failure to understand these semantic issues, based on animalistic lack of foresight, results invariably in a great deal of unnecessary suffering. There is little doubt that without these delusions and illusions we should look after the conditions of our actual lives more closely, and many of our pressing needs would be adjusted.

The difficulties which we have are mostly man-made, and so only mankind can remedy them, and any attempts to escape from *m.o* reality only aggravate the situation.

11. ON THE FOUNDATION OF PSYCHOPHYSIOLOGY

General Considerations

Some of the most important researches in the function of the higher nervous centres have been done lately by Professor Pavlov in his work on the so-called 'conditioned reflexes'. This work was developed in a series of papers covering a period of nearly thirty years of experimentation, but the average international scientist did not know this work as an entirety, because the papers were scattered and written mostly in Russian. Only in 1927 did the Oxford Press publish Pavlov's *Conditioned Reflexes, an Investigation of the Physiological Activity of the Cerebral Cortex* in the English translation of Doctor G. V. Anrep; and in 1928 The International Publishers (New York) published Pavlov's *Lectures on Conditioned Reflexes, Twenty-five Years of Objective Study of the Higher Nervous Activity (Behaviour) of Animals* in the translation of Doctor W. Horsley Gantt. Both translators were collaborators of Professor Pavlov in Leningrad for a number of years. In these two books, the latest experiments and interpretations are given.

Hitherto, most of the researches on the function of the higher nervous systems were formulated in 'psychological' languages, which, obviously, are not fit for physiological disciplines. Professor Pavlov, himself, suggests this fact as an explanation why, until his work, the physiology of the cerebral cortex was so little known. There is no doubt that the descriptive physiological language of happenings, functionings., used by him exclusively, is responsible for his results. This language suggests structurally new experimentations, which suggestions are lacking in other accounts of the kind where antiquated 'psychological' terms are used.

Although I knew as much as the average scientist about the work of Pavlov, this knowledge was not integrated enough to make some issues clear. But, after I had formulated my [non-A]-system, I read the books of Pavlov and found, to my great satisfaction, that a neurological mechanism, the analysis of which underlies my own work, and the existence of which was independently discovered by me on *theoretical* grounds, had been discovered by Professor Pavlov and his co-workers on *experimental* grounds, thus supplying additional experimental verification for my system.

It seems that the so-called 'ethics'., in general, sanity, which underlie desirable human characteristics have a definite *physiological* mechanism, automatically involving on psycho-logical levels these desirable semantic attitudes. It appears that some of the psycho-logical problems enormously complex and difficult to reach, or even inaccessible, are solved, not by preaching, but by the most simple and elementary *physiological* training, a fact which has been verified empirically. Obviously, such simplification, if at all possible, must be of fundamental importance.

Physiology deals, in the main, with the functioning of organs in organisms, and results in various formulations. Thus, there might be an hypothetical 'physiological theory of most effective feeding', for instance, stating that food should be secured first in one's hand, or spoon, or fork, before putting it in the mouth, . A group of people who habitually disregarded the 'physiological theory' and abandoned attempts to act in accordance with it after the first unsuccessful one, would be badly underfed or would simply perish. Facts of experience show that some such 'physiological theory' must have been known and applied from time immemorial, and that, perhaps, because of it we survive at all!

How about the 'mental' field? As I demonstrate—and close observation will verify this very generally—the existing theories of 'mental' life, closely related with our linguistic habits, are

[A], grossly inadequate, and lead to a wholesale production of morons, imbeciles, 'emotionally' disturbed, and, in general, un-sane individuals. Investigation shows the possibility of a simple and obvious *physiological* theory of the use of our nervous system, which automatically leads to desirable psycho-logical, semantic states of general sanity.

In the frivolous example of a 'physiological theory of feeding' given above, the problems of *order* were important. In the physiological theory of sanity, order becomes paramount. Processes and function involve series of states, by necessity exhibiting order. Adjustment to life-conditions means adjustment of processes, and a physiological theory of sanity must be based structurally on four-dimensional order, where 'space' and 'time' are indivisibly interwoven.

Pavlov shows, in an unusually impressive variety and numbers of experiments, how 'order' and 'delay' (four-dimensional order, in the language used here in this connection) are intimately related with most fundamental processes in the higher nervous centres, and how, by the changes or interplays of them, we can produce or eliminate *pathological states* of the nervous system.

In the human field we find a quite similar situation, unanalysable by older methods, because all order involves asymmetrical relations, which, as we have already shown, cannot be dealt with by [A] means.

The issue is clear and definite: either we persist in our old [A] habits of speech, in which case asymmetrical relations and order evade our grasp, and proper evaluation and sanity are *physiologically* impossible, or we build a [non-A]-system free, or at least more free, from these evaluational limitations, which allows us to deal with order, and sanity becomes *physiologically possible*.

'Stimuli are never "simple", and, by necessity, involve fourfold space-time structure and order. Survival values involve, also, this four-dimensional order. For instance, the natural survival order is "senses" first, "mind" next; object first, label next; description first, inference next , . The reversal of the natural order appears pathological and pathogenic and is found as a symptom in practically all forms of "mental" ills, as well as in most human difficulties and disturbances which, at present, are still not considered abnormal. Thus, objectivity is ascribed to words, "mind" projected into "senses", inferences evaluated as descriptions., —quite common "symptoms" . . . Observations on human levels show that we still copy animals in our nervous responses, confuse orders of abstractions (non-existent for the animals), leading fatalistically to the reversal of the natural order and to pathological results, making the great majority of us un-sane.' (From Discussion by A. Korzybski. *Proceedings of the First International Congress of Mental Hygiene.* New York, 1932.)

A structural *non-el* enquiry into the objective world shows quite clearly that no event is ever 'simple'; it is, at least, a limited whole of interrelated factors. The eventual 'simplicity' is manufactured by a nervous process of higher and higher abstractions.

In our consideration of 'order' and 'delay', and the role they play in connection with the activities of the nervous system, we must first discriminate sharply between the objective level which is *un-speakable*, because anything that can be said *is not* the object, and the verbal level, on which we can, at will, concentrate attention on similarities, or differences, or both. Secondly, we must pay special attention to structure—that is to say, search for structure in the empirical world, and, once this has been found, adjust, accordingly, the structure of our language.

The structure of the daily, as well as of the 'philosophical', language, which we inherited, in the main, from our primitive ancestors, is such that we have *separate terms* for factors which are not separable, such as 'matter', 'space', 'time', or 'body', 'soul', 'mind', . Then, as it were, we try to make out of the word, flesh, by reversing the natural order and affectively ascribing a delusional *objectivity* to these terms.

If we deal with the silent, un-speakable, objective level and try to divide according to the implications of the verbal division, we find a brutal fact, which, until Einstein and Minkowski, has escaped scientific verbal formulation, that this cannot be done at all. On the objective level every dealing with 'matter' involves 'space' and 'time'; any dealing with 'space' involves some fulness of something and 'time'; and every dealing with 'time' involves 'something' and 'space'.

The structure of the world happens to be such that empirically 'matter', 'space', 'time', cannot be divided; wherefore, we should have a *non-el* language of *similar structure*. This was accomplished by Einstein-Minkowski, when they created a language of 'space-time', in which the hard lumps against which we bump our noses are connected analytically with the curvature of space-time.

In this new, *non-el*, four-dimensional language, every three-dimensional point of 'space' has a date, and so is different. For our purpose, we do not need, at present, to bother much about its curvature or the kinks in space-time, called in the old way 'matter', but we must emphasize that the fourfold order is of great importance, as it corresponds structurally to *experience*, and is intimately connected with physiological reactions, the semantic included.

On Conditional Reflexes

In our field, where we have to formulate sharp differences between the nervous responses of 'man' and 'animal', we say that animals stop abstracting or linking of signals on some level, while humans do not. The latter abstract in indefinitely higher orders—at least potentially.

Here we encounter a fundamental and sharp far-reaching difference between the nervous functioning of 'animal' and 'man'. This abstracting in indefinitely higher orders no doubt conditions the mechanism of what we call human 'mentality'. If we stop this abstracting anywhere, and rest content with it, we copy animals in our nervous processes, involving animalistic *s.r.* As will be shown later, this is the actual case with practically all of us, owing to our [A] education and theories. This 'copying animals' in our nervous responses is, perhaps, a natural tendency at an extremely low level of development; but as soon as we understand the physiological mechanism, we can correct our education, with corresponding human semantic results. Naturally, such 'copying animals' by humans must be a process of arrested development or regression. It must be pathological for man, no matter how severe or how mild the affliction may be. Various absolutists, and the 'mentally' ill in general, show this semantic mechanism clearly.

The reactions can be divided into two groups, those which are *inborn*, almost automatic, almost unconditional, rather few and simple, belonging to the so-called 'species'; and those which are *acquired* during individual life, allow a great variety of complications, are *conditional in different degrees*, and are acquired by the individual. Pavlov suggests different terminologies; for instance, he calls the one 'inborn', the other 'acquired'; or as usually incorrectly translated into English as 'unconditioned' and 'conditioned' respectively. The two last terms have received a scientific general acceptance, yet I would suggest that in the English incorrect translation they are *structurally unsatisfactory*, and that particularly, when applied to humans, they carry harmful implications. Structurally, 'inborn' and 'acquired' are entirely satisfactory. Terms like 'conditional' and 'unconditional' (in the original language of Pavlov), although less satisfactory, are more appropriate, as they do not imply some sort of 'cause-lessness'. In fact, the 'unconditioned' salivary reactions *are conditioned* and produced by the physico-chemical effect of the food, and so to call them 'unconditioned' is structurally erroneous. The terms 'conditional' and 'unconditional' do not have similar implications, and carry others, as, for instance, the possibility of very important *degrees of conditionality*, establishing the ∞-valued character of the reactions; conditional meaning non-absolute, and non-one-valued.

For these structural reasons, I shall use the terms 'inborn' and 'acquired' or else 'unconditional' and 'conditional' reactions.

Under natural conditions, an animal, to survive, must respond not only to normal stimuli, which bring immediate harm or benefit, but also to different physical and chemical stimuli, in themselves neutral, such as waves of sound or light., which are *signals* for animals and *symbols* for man. The number of inborn reactions is comparatively small, and, alone, they are not sufficient for the survival of higher animals in their more complex environment. Experiments have made this point quite obvious. A completely decorticated animal may retain his inborn reactions and become a kind of automatic mechanism; but all his subtler means of adjustment, owing to acquired reactions, disappear, and if unaided he can not survive. Thus, a decorticated dog will only eat when food is introduced into his mouth, and would otherwise die of starvation though food be placed all around him.

Experimental evidence seems to show that all higher activities of the nervous system, the whole signalizing apparatus, which underlies the formation and maintenance of the acquired conditional reactions, depend on the integrity of the cortex. Stimuli which produce conditional reactions are acting as signals of benefit or danger. These signals are sometimes nominally 'simple', sometimes very complex, and the structure of the nervous system is such that it can abstract, analyse, and synthetize the factors of importance for the organism, and integrate them into excitatory complexes. The analysing and synthetizing functions, as usual, overlap, and cannot be sharply divided, both functions being only aspects of the manifestation of the activity of the nervous system as-a-whole. In general, one of the most important functions of the cerebral cortex is that of reacting to innumerable stimuli of variable significance, which act as signals in animals and symbols in humans, and give means of very subtle adjustment of the organism to the environment. In psycho-logical terms, we speak of 'associations', 'selection', 'intelligence'.; in mathematical terms, of relations, structure, order.; in psychophysiological terms, of semantic reactions.

The language of reactions is of special interest because its structure is similar to the structure of protoplasm in general and the nervous system in particular. This language can be expanded and supplemented by the following further structural observations:

1) That reactions in animals and humans exhibit *different degrees of conditionality;*

2) That the signals and symbols may have *different orders*, indicating superimposition of stimuli;

3) That animals cannot extend their responses to signals of higher order indefinitely;

4) That humans can extend their semantic responses to higher order symbols indefinitely, and, in fact, have done so through language which is always connected with *some* response, be it only repression or some other neurotic or psychotic manifestations.

The above extension is structurally fundamental, because we can extend the vocabulary of conditional reactions to humans in all their functions. Without it, we find ourselves saddled with a vocabulary which does not correspond in structure to the well-known elementary facts concerning *human* responses to stimuli, and we relapse into the old 'behaviourism', which is structurally insufficient.

The present system is based on such observations and extensions. It was reached independently from structural and physico-mathematical considerations. With this structural verbal extension, we can easily be convinced that everything that we call 'education', 'habits', 'learning'., on all levels is building up acquired or conditional and s.r of *different orders*, as one of the differences between 'man' and 'animal' consists in the fact that humans can extend their symbolism and responses to indefinitely high orders, while with animals this power of abstracting and response *stops somewhere*. We establish here a sharp distinction between the high abstractions 'man' and 'animal', and so build up a psychophysiological and structurally satisfactory language.

In interpreting the experiments on animals as applied to humans, it should be remembered that some of the experiments of Pavlov, *as they stand*, would be, at the least, *neurotic* for man. The reason for this is that the higher abstractions of man, which are due to the more developed complexities of his nervous system, would often make such simple experiments impossible. Once a conditional reaction is established with an animal, no amount of any sort of 'intellectual' persuasion, or the like, would disturb his glandular secretions, as the animal's range of 'meanings' is very limited. These secretions can be diminished or even abolished by other means, but not by 'intellectual' means alone. In the 'normal' man, his 'knowing' that the sound of the metronome or bell is part of an experiment and not a signal for actual food, would, *or should*, alter his nervous reactions and glandular secretions and make the experiments much more complex. The conditional reactions of the animals have still the *element of unconditionality*. In man, they may become *fully conditional* and depend on a much larger number of semantic factors called 'mental', 'psychic'., than we find in any animal.

On the human level, outside of the experiments with the salivary glands, we have in the psychogalvanic reaction a most subtle semantic means of experimenting with the effect of words as connected with some secretions, probably at least the sweat glands. Humans react to different events or words by minute electrical currents (among others) which can be registered by a very sensitive galvanometer and the curves photographed. It is interesting to notice that so-called 'self-consciousness' disturbs the success of the experiments, or makes them impossible, *at least with some individuals*. It should be remembered that general statements are invalidated if there are any exceptions.

In experiments, we are usually interested in their success. When analysing the ∞-valued *degrees of conditionality*, we are equally interested in their failures, which suggest a far-reaching revision of the *interpretation* of our experimental data in this field. Although some writers say that the reactions registered are 'beyond control' (unconditional), this statement, in general, is not correct, and should be amended to 'often beyond control' (conditional of different degrees). It is impossible to go into details here, as the problems are extremely complex. In addition to this, the testing of *degrees of conditionality* presents an extremely wide *new semantic field* for experimentation which has not yet been attempted. It should be noted, however, in passing, that in these experiments different types of 'mentally' ill, as well as the 'healthy' persons, exhibit different types of curves.

When psycho-logical events or s.r are interpreted, the difficulties become particularly acute. Thus, we seldom discriminate between the average and the 'normal' person. In the *animal world*, under natural conditions—by which is meant entirely without human interference—the survival conditions are *two-valued* and very sharp. The animals survive or they die out. Because of this, it could be said, with regard to the animal world, with some sort of plausibility, that the average, with a long list of specifications, could be considered the 'normal' animal. We usually enlarge this notion to humans and land in fallacies, particularly in so-called 'psychological' problems, which admittedly are very difficult.

In general medical science, such mistakes are made more seldom. No physician, studying a colony of lepers or syphilitics, could conclude that a 'normal' man should be a leper or a syphilitic. He would say that probably, in a given colony, the average person is afflicted with such and such a disease, and he would keep as his medical standard for desirable health, a 'normal' man; that is to say, one free from this disease.

It is true that in the example given above, outside of such rare colonies, we have a majority which, in respect to the given disease, are healthy; so we are empirically forewarned against fallacies, although existing theories of knowledge do not forewarn us. But the main point remains true; namely, that in human life the average 1933 does not mean 'normal', and the standard for 'normal' will have to be established *exclusively* by scientific research. In our present work, we show that the average person copies animals in his psycho-logical and nervous processes, exhibits the unconditionality of nervous responses, confuses orders of abstractions, reverses the natural order., semantic symptoms of similar *structure* as found in obviously 'mentally' ill. Therefore, the average person 1933 *must be considered pathological*. If we take the animalistic average for 'normal', and apply it to man, we commit a similar fallacy as that of treating a colony of lepers as a 'normal' or 'healthy' group.

In conditional *s.r* of man, the average person cultivates, through inheritance and training in [A] doctrines, languages of inappropriate structure., animalistic, nervous, and so psycho-logical, *s.r.* But here, as in general medicine, the average pathological situation should not be considered 'normal'. Only a structural study can disclose what with man should be considered 'normal'. The present system performs this task to a limited degree and in various ways, among others, by the revision and the widening of the reaction vocabulary to a larger structural conditionality, as found in the, as yet, exceptional 'normal' man, and introduces the important notion of *non-elementalistic semantic reactions*.

Because of this 'average for normal' fallacy, the theories of 'conditional reflexes' in man should be thoroughly revised and enlarged to include *non-elementalistic semantic reactions;* and then we should find that often what is 'normal' with animals is quite pathological for man. The semantic difficulties are serious, because the accepted two-valued structure of language and semantic habits reflect the primitive mythologies; so there is always the danger of drifting either into animalism, or into some other sort of equally primitive mysticism.

The net psycho-logical result of such a revision appears to be that, on structural grounds, what on human level appears as desirable, and, at present, exceptional—as, for instance, the complete conditionality of conditional and *s.r*, based on the consciousness of abstracting—ought to be considered the rule for a 'normal' man. Then the older animalistic generalizations will become invalid and reactions transformed. But for this purpose, and to be able to apply these considerations in practice, we shall have to analyse 'consciousness of abstracting' and, therefore, 'consciousness' which must be defined in simpler terms, discussed in Part VII.

When we deal with 'mentally' ill persons, the reactions which would be conditional with 'normal' persons become, in a sense, unconditional, compulsory, and semi-automatic in effect, inwardly as well as outwardly. As with animals, no amount of 'intellectual' persuasion has any effect on them, and the reactions, secretions., follow automatically. From the physiological point of view, 'mental' ills in humans compare well with *conditional reactions in animals*. It seems that under such circumstances a physiological language of different orders of abstractions, different orders of conditional and *s.r* would be structurally satisfactory. In such a language, we should pass from the inborn reactions, which exhibit the maximum of persistence, unconditionality, and almost automatic character, to the acquired or conditional reactions *in animals*, which would be called *lower order conditional reactions*, still, to some extent, automatic in their working, and, finally, to the much more flexible, variable, ∞-valued and *potentially fully conditional reactions in man*, which we will call *conditional reactions of higher orders* which include the *semantic reactions*.

In such a vocabulary, the main term 'reaction' would be retained as a structural implication; yet the *degrees of conditionality* would be established by the terms of 'lower order' or 'higher order' conditional reactions. Such a language would have the enormous advantage of being physiological and ∞-valued. Structurally, it would be in accordance with what we know from psychiatry; namely, that the 'mentally' ill exhibit arrested development or regressive tendencies.

We would say that 'mental' illness exhibits not only arrested development or regression, but we could state definitely that the *fully conditional* (∞-valued) reactions of higher order have not developed enough, or have degenerated (regression) into *less conditional* (few-valued) reactions of lower orders as found in animals. All the 'phobias', 'panics', 'compulsory actions', identifications or confusions of orders of abstractions., show a similar semantic mechanism of mis-evaluation. Although they naturally belong to the so-called 'conditional reactions', yet, being impervious to reason, they have the one-valued character of *unconditionality*, as in animals.

Similarly with the difference between signals and symbols. The signal with the animal is *less* conditional, more one-valued, 'absolute', and involves the animal in the responses which we have named conditional reactions of lower order. Symbols with the *normally developed man* (see discussion of 'normal' above) are, or should be, ∞-valued, indefinitely conditional, not automatic; the *meanings*, and, therefore, the situation as-a-whole, or the context in a given case, become paramount, and the reactions should be fully conditional—that is to say, reactions of higher order. In human regression or undevelopment, human symbols have degenerated to the value of signals effective with animals, the main difference being in the *degree* of conditionality. Absolutism as a semantic tendency in humans involves, of necessity, one- or few-valued attitudes, the lack of conditionality, and thus represents a pre-human tendency.

To what extent the language of the *degrees of conditionality* is helpful in understanding the development of *human* 'intelligence', and why a fully developed *human* 'mind' should be related with *fully conditional* reactions of higher order, can be well illustrated by an example taken quite low in the scale of life.

This example is selected only because it is simple, and illustrates an important principle very clearly. We know that fishes have a well-developed nervous system, do not possess a differentiated cerebral cortex; but experiments show that they can learn by experience. If we take a pike (or a perch) and put it in a tank in which some minnows, its natural food, are separated from it by a glass partition, the pike will dash repeatedly against the glass partition to capture the minnows. After a number of such dashes it abandons the attempt. If we then remove the partition, the pike and the minnows will freely swim together and the pike will not attempt to capture the minnows.

The dash for capturing the minnows was a positive and unconditional, inborn feeding reaction, unsuited for the environmental conditions as they happened to be at that moment. The (perhaps) painful striking of the glass was a negative stimulus, which abolished the positive reaction—speaking descriptively—and established a negative conditional reaction, the result of individual experience, which, as we observe by the actions of the fish, is not flexible, not adjustable, and quite rigid, one-valued, and semi-unconditional, or of low degree of conditionality, because, when the glass partition is removed, the pike swims freely among the minnows without adjusting itself to the new conditions and capturing the minnows.

A cat separated from a mouse by a glass partition also stops dashing against the glass, but this negative reaction is *more conditional*. In 'psychological' terms, the cat is 'more intelligent', *evaluates relations* better than the fish, and when the glass partition is removed, the cat captures the mouse almost immediately.

In this connection, an interesting experiment could be made, though I am not aware that it has been performed; namely, to separate the above fishes with a wire screen, which would be *visible* to the fishes, and repeat the experiments to test if the removal of a *visible* obstacle would alter the outcome of the experiment or the 'time' of the reactions. If the 'time' for capturing the minnows were reduced, this would mean that the conditionality of the reaction was increased, and so the seeing of the obstacle, or the increased power of abstraction, would play some role in it. Even humans are deceived by Houdinis. Are we so 'superior' to the 'poor fish'?

In the process of human evolution from the lowest savage to the highest civilized man, it is natural that we should pass through a period in which the primitive doctrines and languages must be revised. The newest achievements in science indicate that the twentieth century may be such a period. Even in mathematics and physics, to say nothing of other disciplines, it is only the other day that the old elementalism and two-valued semantics were abandoned. Obviously, consciousness of abstracting produces *complete conditionality* in our conditional higher order reactions, and so must be the foundation on which a science of man, or a theory of sanity and human progress, must be built.

The suggested extension of the reaction vocabulary would allow us, at least, to apply a uniform physiological language to life, *man included*. We should have a general language for life and all activities, 'mind' included, of a structure similar to the known protoplasmic and nervous structure, not excepting the highest activities. 'Mental' ills would be considered as arrested development or regression to one-, or few-valued semantic levels; sanity would be in the other direction; namely, progression conditioned by larger and larger flexibility of conditional and semantic reactions of higher order, which, through ∞-valued semantics, would help adjustment under the most complex social and economic conditions for man. The maximum of conditionality would be reached, let us repeat, through the consciousness of abstracting, which is fundamental for sanity, and is the main object of the present work, explained in Part VII.

12. ON ABSTRACTING

Aristotle, in building his theories, had at his disposal, besides his personal gifts, a good education according to his day and the science current in 400-300 B.C. Even in those days, the Greek language was a very elaborate affair. Aristotle and his followers simply took this language for granted. The problems of the structure of language and its effect on *s.r* had not yet arisen. To them, the language they used was *the* (unique) language. When I use the expression '*the* language', I do not mean anything connected with the language, as *Greek;* I mean only the structure of it, which was much similar in the other national languages of this group. The language Aristotle inherited was of great antiquity, and originated in periods when knowledge was still more scanty. Being a keen observer, and scientifically and methodologically inclined, he took this language for granted and systematized the modes of speaking. This systematization was called 'logic'. The primitive structural metaphysics underlying this inherited language, and expressed in its structure, became also the 'philosophical' background of this system. The subject-predicate form, the 'is' of identity, and the elementalism of the [A]-system are perhaps the main semantic factors in need of revision, as they are found to be the foundation of the insufficiency of this system and represent the mechanism of semantic disturbances, making general adjustment and sanity impossible. These doctrines have come down to us, and through the mechanism of language the semantic disturbing factors are forced upon our children. A whole procedure of training in delusional values was thus started for future generations.

As the work of Aristotle was, at his date, the most advanced and 'scientific', quite naturally its influence was wide-spread. In those days, no one spoke of this influence as 'linguistic', involving *s.r.* Aristotle's work was, and still is, spoken of as 'philosophy', and we speak mostly of the influence of [A] 'philosophy' rather than of the [A] structure of language, and its semantic influence.

As we have already seen, when we make any proposition whatsoever we involve creeds, or metaphysics, which are embodied silently as structural assumptions and in our undefined terms. The use of terms notdefinable in simpler terms at a given date is inherent and seemingly unavoidable.

When our primitive ancestors were building their language, quite naturally they started with the lowest orders of abstractions, which are the most immediately connected with the outside world. They established a language of 'sensations'. Like infants, they identified their feelings with the outside world and personified most of the outside events.

This primitive semantic tendency resulted in the building of a language in which the 'is' of identity was fundamental. If we saw an animal and called it 'dog' and saw another animal roughly resembling the first, we said, quite happily, 'it *is* a dog', forgetting or not knowing that the objective level is un-speakable and that we deal only with absolute individuals, each one different from the other. Thus the mechanism of identification or confusion of orders of abstractions, natural at a very primitive stage of human development, became systematized and structurally embodied in this most important tool of daily use called 'language'. Having to deal with many *objects*, they had to have names for objects. These names were 'substantives'. They built 'substantives', grammatically speaking, for other feelings which were not 'substantives', ('colour', 'heat', 'soul', .). Judging by the lower order abstractions, they built adjectives and made a completely anthropomorphised world-picture. Speaking about speaking, let us be perfectly aware from the beginning that, when we make the simplest statement of any sort, this statement already presupposes some kind of structural metaphysics. The early vague feelings and savage speculations about the structure of this world, based on primitive insufficient scientific data, was influencing the building of the language. Once the language was built, and, particularly, systematized, these primitive structural metaphysics and *s.r* had to be projected or reflected on the outside world—a procedure which became habitual and automatic.

Was such a language structurally reliable and safe? If we investigate, we can easily become convinced that it was not. Let us take three pails of water; the first at the temperature of 10° centigrade, the second at 30°, and the third at 50°. Let us put the left hand in the first pail and the right in the third. If we presently withdraw the left hand from the first pail and put it in the second, we feel how nicely *warm* the water in the second pail is. But, if we withdraw the right hand from the third pail and put it in the second, we notice how *cold* the water is. The temperature of the water in the second pail was practically not different in the two cases, yet our feelings registered a marked difference. The difference in the 'feel' depended on the former conditions to which our hands had been subjected. Thus, we see that a language of 'senses' is not a very reliable language, and that we cannot depend on it for general purposes of evaluation.

How about the term 'dog'? The number of individuals with which any one is directly acquainted is, by necessity, limited, and usually is small. Let us imagine that someone had dealt only with good-natured 'dogs', and had never been bitten by any of them. Next he sees some animal; he says, 'This *is* a dog'; his associations (relations) do not suggest a bite; he approaches the animal and begins to play with him, and is bitten. Was the statement 'this *is* a dog' a safe statement? Obviously not. He approached the animal with semantic expectations and *evaluation* of his verbal definition, but was bitten by the non-verbal, un-speakable objective level, which has different characteristics.

Judging by present standards, knowledge in the days of Aristotle was very meagre. It was comparatively easy 2300 years ago to summarize the few facts known, and so to build generalizations which would cover those few facts.

If we attempt to build a [non-A]-system, 1933, can we escape the difficulties which beset Aristotle? The answer is that some difficulties are avoidable, but that some are inherent in the structure of human knowledge, and so cannot be entirely evaded. We can, however, invent new methods by which the harmful semantic effect of these limitations can be successfully eliminated.

There is no escape from the fact that we must start with undefined terms which express silent, structural creeds or metaphysics. If we state our undefined terms explicitly, we, at least, make our metaphysics conscious and public, and so we facilitate criticism, co-operation, . The modern undefined scientific terms, such as 'order', for instance, underlie the exact sciences and our wider world-outlook. We must start with these undefined terms as well as the modern structural world-outlook as given by science, 1933. That settles the important semantic point of our structural metaphysics. It need hardly be emphasized that in a human class of life, where creeds are characterized by having dates, they *should* always be labeled with this date. For sanity, the creeds utilized in 1933 should be of the issue of 1933.

Now as to the *structure* of our language. What structure shall we give to our language? Shall we keep the old structure, with all its primitive implications and corresponding *s.r*, or shall we deliberately build a language of new structure which will carry new modern implications and *s.r*? There seems to be only one reasonable choice. For a [non-A]-system, we must build a new language. We must abandon the 'is' of identity, to say the least. We have already seen that we have an excellent substitute in an actional, behaviouristic, operational, functional language. This type of language involves modern asymmetrical implications of 'order', and eliminates the 'is' of identity, which always introduces false evaluation.

To these fundamental starting points, we must add the principle that our language should be of *non-el* structure. With these *minimum* semantic requirements, we are ready to proceed.

Let us take any object of ordinary experience, let us say the one we usually call a 'pencil', and let us briefly analyse our nervous relationship to it. We can see it, touch it, smell it, taste it., and use it in different ways. Is any of the relationships just mentioned an 'all-embracing' one, or is our acquaintance through any of them only *partial*? Obviously, each of these means provides an acquaintance with this object which is not only *partial*, but is also *specific* for the nerve centres which are engaged. Thus, when we look at the object, we do not get odor or taste stimuli, but only visual stimuli, .

If the object we call 'pencil' were lying on the surface of this paper and we were to look at it along the surface of the paper in a perpendicular direction to its length, it would generally be seen as an elongated object, pointed at one end. But, if we were to observe it along the plane of the paper at right angles to our former direction, it would be seen as a disk. This illustration is rough, but serves to show that the acquaintance derived through any specific means (e.g., vision) is also *partial* in another sense; it varies with the position., of any specified observer, Smith, or a camera.

Furthermore, any given means provides, for *different* observers, different acquaintances. Thus, vision shows the pencil to one observer, Smith, as a pointed rod, and to another observer, Jones, as a disk. Feeling, through other receptors, is just as dependent upon many conditions; and different observers receive different impressions. This is well illustrated by the familiar tale of the five blind men and the elephant.

Because of differences in sensitivity in the receptors of Smiths and Browns (partial colour-blindness, astigmatism, far-sightedness.,), any given means of acquaintance (e.g., vision) gives to different observers different reports of the one object. The acquaintance is thus personal and individual.

Again, the reports received through particular channels are influenced by the kind of reports that have already come through that channel. To one who has not seen trees frequently, a spruce and a balsam are not seen to be different. They are just 'evergreens'. With better educated seeing, this individual later differentiates, perhaps, four kinds of spruce. Because of this factor of experience, the *response* of each individual to similar external stimuli is individual. We can only *agree* on colours, shapes, distances., by ignoring the fact that the effect of the 'same' stimulus is different in different individuals. Besides that, we have no accurate means of comparing our impressions.

The 'time' factor enters, in that we cannot become acquainted with our pencil *on all sides at once*. Nor can we observe the outer form and the inner structure at the 'same time'. We may even neglect to examine the inner structure entirely. Even more important is the fact that all our means together give us only a *partial* and personal acquaintance with the 'pencil'. Continually we invent extra-neural means which reveal new characteristics and finer detail. Nor is this process ever completed. No one can ever acquire a 'complete' acquaintance with even so simple an object as a pencil. The chemistry, the physics, the uses of the varieties., offer fields of acquaintance that can be extended indefinitely. Nature is inexhaustible; the events have infinite numbers of characteristics, and this accounts for the wealth and infinite numbers of possibilities in nature.

I used the word 'acquaintance' deliberately, because it seems vague, and, as yet, *el* gambling on words have not spoiled this term. I had to avoid the *el* terms 'senses' and 'mind' as much as possible in this analysis. If we recall the example of paper roses in the case of hay fever, we shall realize that the terms 'senses' and 'mind' are not reliable, particularly in humans. As a further instance, we have but to remember the experiment with newspaper headlines, also cited earlier.

We become better acquainted with the object by exploring it in manifold ways, and building for ourselves different pictures, all partial, and supplied by direct or indirect contact with different nerve centres. In these explorations, different nerve centres supply their *specific* responses to the different stimuli. Other higher nerve centres summarize them, eliminate weaker details, and so, gradually, our acquaintance becomes fuller while yet remaining *specific* and *partial,* and the semantic problems of *evaluation, meanings,* begin to be important.

If we try to select a term which would describe structurally the processes which are essential for our acquaintance with the object, we should select a term which implies 'non-allness' and the specificity of the response to the stimuli.

If we pass from such a primitive level to a level of 1933, and enquire what we actually know about an object and the structure of its material, we find that in 1933 we know positively that the internal structure of materials is very *different* from what we gather by our rough 'senses' on the macroscopic level. It appears of a dynamic character and of an extremely fine structure, which neither light, nor the nerve centres affected by light, can register.

What we see is structurally only a specific *statistical mass-effect* of happenings on a much finer grained level. We *see* what we see because we *miss* all the finer details. For our purpose, it is usually enough to deal only with sight; this simplifies writing, and the comments made apply to all other 'senses', though perhaps in different degrees.

In 1933, in our human economy, we have to take into account at least three levels. The one is the sub-microscopic level of science, what science 'knows' *about* 'it'. The second is the gross macroscopic, daily experience level of rough objects. The third is the verbal level.

We must also evaluate an important semantic issue; namely, the relative importance of these three levels. We know already that to become acquainted with an object, we must not only explore it from all possible points of view and put it in contact with as many nerve centres as we can, as this is an essential condition of 'knowing', but we must also not forget that our nerve centres must summarize the different partial, abstracted, specific pictures. In the human class of life, we find a new factor, non-existent in any other form of life; namely, that we have a capacity to collect all known experiences of different individuals. Such a capacity increases enormously the number of observations a single individual can handle, and so our acquaintance with the world around, and in, us becomes much more refined and exact. This capacity, which I call the time-binding capacity, is only possible because, in distinction from the animals, we have evolved, or perfected, extra-neural means by which, without altering our nervous system, we can refine its operation and expand its scope. Our scientific instruments record what ordinarily we cannot see, hear, . Our neural verbal centres allow us to exchange and accumulate experiences, although no one could live through all of them; and they would be soon forgotten if we had no neural and extraneural means to record them.

Again the organism works as-a-whole. All forms of human activities are interconnected. It is impossible to select a special characteristic and treat it in a delusional *el* 'isolation' as the most important. Science becomes an extra-neural extension of the *human* nervous system. We might expect the structure of the nervous system to throw some light on the structure of science; and, vice versa, the structure of science might elucidate the working of the human nervous system.

This fact is very important, semantically, and usually is not sufficiently emphasized or analysed enough. When we take these undeniable facts into account, we find the results already reached to be quite natural and necessary, and we understand better why an individual cannot be considered entirely sane if he is wholly ignorant of *scientific method* and structure, and so retains primitive *s.r.*

For a theory of sanity, all three levels are important. Our 'senses' react as they do because they are united as-a-whole in one living structure, which has potentialities or capacities for language and science.

If we enquire what we *do* in science, we find that we 'observe' silently and then record our observations *verbally*. From a neurological point of view, we abstract whatever we and the instruments can; then we summarize; and, finally, we generalize, by which we mean the processes of abstracting carried further.

In our 'acquaintance' with daily objects, we do substantially a similar thing. We abstract whatever we can, and, according to the degree of intelligence and information we have, we summarize and generalize. From the psychophysiological point of view, the ignorant is neurologically deficient. But to 'know' or to 'believe' something which is false to facts is still more dangerous and akin to delusions, as psychiatry and daily experience teach us. It is a neurological fallacy to treat science in 'isolation' and disregard its psychophysiological role.

In the building of our language, a similar neurological process becomes evident. If we were to see a series of different individuals, whom we might call Smith, Brown, Jones., we could, by a process of abstracting the characteristics, segregate the individuals by sizes or colours.; then, by concentration on one characteristic and disregarding the others, we could build classes or higher abstractions, such as 'whites', 'blacks', . Abstracting again, with rejection of the colour difference., we would finally reach the term 'man'. This procedure is general.

Anthropological studies show clearly how the degree of 'culture' among primitive peoples can be measured by the orders of the abstractions they have produced. Primitive languages are characterized particularly by an enormous number of names for individual objects. Some savage races have names for a pine or an oak., but have no 'tree', which is a higher abstraction from 'pines', 'oaks', . Some other tribes have the term 'tree', but do not have a still higher abstraction 'woods'. It does not need much emphasis to see that higher abstractions are extremely *expedient* devices. There is an enormous economy which facilitates mutual understanding in being able to be brief in a statement and yet cover wider subjects.

Let us consider a primitive statement 'I have seen tree$_1$', followed by a description of the individual characteristics 'I have seen tree$_2$', with minute individual description., where tree$_1$, tree$_2$., stand for names of the individual trees. If an event of interest had happened in a place where there were a hundred trees, it would take a long while to observe fairly well the individual trees and still longer to give an approximate description of them. Such a method is non-expedient, *fundamentally endless;* the mechanism is cumbersome, involves many *irrelevant* characteristics; and it is impossible to express in a few words much that might be *important*. Progress must be slow; the general level of development of a given race or individual must be low. It should be noticed that the problem of *evaluation* enters, at once implying many most important psycho-logical and semantic processes. Similar remarks apply to the abstracting of infants, 'mentally' deficient grown-ups, and some 'mentally' ill.

Indeed, as the readers of my *Manhood of Humanity* already know, the 'human class of life' is chiefly differentiated from 'animals' by its rapid rate of progress through the rapid rate of accumulation of past experiences. This is possible only when expedient means of communication are established; that is, when higher and higher orders of abstractions are worked out.

All scientific 'laws', and other generalizations of higher order (even single words), are precisely such methods of expediency, and represent abstractions of very high order. They are uniquely important because they accelerate progress and help the further summarizing and abstracting of results achieved by others. Naturally, this process of abstracting has also unique practical consequences. When chemical 'elements' were 'permanent' and 'immutable', our physics and chemistry were much undeveloped. With the advent of higher abstractions, such as the monistic and general dynamic theories of all 'matter' and 'electricity', unitary field theories., the creative freedom of science and the control over 'nature' have increased enormously and will increase still more.

Psychiatry also seems to give data indicating that 'mental' illnesses are connected either with arrested development or with regression to phylogenetically older and more primitive levels, all of which, of course, involves lower order abstractions. From the point of view of a theory of sanity, a sharp differentiation between 'man' and 'animal' becomes imperative. For with 'man', the lack of knowledge of this difference may lead to the copying of animals, which would involve semantic *regression* and ultimately become a 'mental' illness.

Although organisms have had acquaintance with objects for many hundreds or thousands of millions of years, the higher abstractions which characterize 'man' are only a few hundreds of thousands of years old. As a result, the nervous currents have a natural tendency to select the older, more travelled, nervous paths. Education should counteract this tendency which, from a *human* point of view, represents regression or under-development.

By now we know how important it is for a [non-A]-system to abandon the older implications and adopt an actional, behaviouristic, operational, or functional language. On the neurological level, what the nervous system *does* is abstracting, of which the summarization, integration., are only special aspects. Hence, I select the term *abstracting* as fundamental.

The standard meaning of 'abstract', 'abstracting' implies 'selecting', 'picking out', 'separating', 'summarizing', 'deducting', 'removing', 'omitting', 'disengaging', 'taking away', 'stripping', and, as an adjective, not 'concrete'. We see that the term 'abstracting' implies structurally and semantically the activities characteristic of the nervous system, and so serves as an excellent *functional physiological* term.

There are other reasons for making the term 'abstracting' fundamental, which, from a *practical* point of view, are important. A bad habit cannot be easily eliminated except by forming a new semantic counter-reaction. All of us have some undesirable but thoroughly established *linguistic habits* and s.r which have become almost automatic, overloaded with unconscious 'emotional' evaluation. This is the reason why new 'non-systems' are, in the beginning, so extremely difficult to acquire. We have to break down the old structural habits before we can acquire the new s.r. The [non-E] geometries or the [non-N] systems are not any more difficult than the older systems were. Perhaps they are even simpler. The main semantic difficulty, for those accustomed to the old, consists in breaking the old structural linguistic habits, in becoming once more flexible and receptive in feelings, and in acquiring new s.r. Similar remarks apply in a more marked degree to a [non-A]-system. The majority of us have very little to do *directly* with [non-E] or [non-N] systems (although indirectly we all have a good deal to do with them). But all of us live our immediate lives in a human world still desperately [A]. Hence a [non-A]-system, no matter what benefits it may give, is much handicapped by the old semantic blockages.

In building such a system, this natural resistance or persistence of the old s.r must be taken into consideration and, if possible, counteracted. One of the most pernicious bad habits which we have acquired 'emotionally' from the old language is the feeling of 'allness', of 'concreteness', in connection with the 'is' of identity and elementalism. One of the main points in the present [non-A]-system is first to remove entirely from our *s.r* this 'allness' and 'concreteness', both of which are structurally unjustified and lead to identification, absolutism, dogmatism, and other semantic disturbances. Usually, the term 'abstract' is contrasted with 'concrete', which is connected with some vague feeling of 'allness'. By making the functional term *abstracting* fundamental, we establish a most efficient semantic counter-reaction to replace the older terms which had such vicious structural implications. Indeed, it is comparatively easy to accept the term 'abstractions of different orders', and any one who does so will see how much clarity and how much semantic balance he will automatically acquire.

From a *non-el* point of view, the term 'abstracting' is also very satisfactory. The structure of the nervous system is in ordered levels, and all levels go through the process of abstracting from the other levels.

The term implies a general activity, not only of the nervous system as-a-whole, but even of all living protoplasm, as already explained. The characteristic activities of the nervous system, such as summarizing, integrating., are also included by implication.

If we wish to use our terms in the strictly *non-el* way, we must abandon the older division of 'physiological abstractions', which implies 'body', and of 'mental abstractions', which, in turn, implies 'mind', both taken in an *el* way. We can easily do that by postulating abstractions of different orders. We should notice that the above use of the term 'abstracting' differs from the old usage. The semantic difference is in uniting all the abstractions our nervous system performs under the one term, and in distinguishing between different abstractions by the order of them, which is functionally, as well as structurally, justified.

The term 'first order abstractions' or 'abstractions of lower order' does not distinguish between 'body' and 'mind'. *Practically*, it corresponds roughly to 'senses' or immediate feelings, except that by implication it *does not eliminate 'mind'*. Neither does the term 'abstractions of higher orders' eliminate 'body' or 'senses', although it corresponds roughly to 'mental' processes.

From the point of view of 'order', the term 'abstracting' has a great deal in its favor. We have seen what serious structural and semantic importance the term 'order' has, and how the activity of the nervous system has to be spoken of in terms of order. If we establish the term 'abstracting' as fundamental for its *general* semantic implications, we can easily make the meanings more definite and specific in each case by having 'abstractions of different orders'.

We have seen also that the terms we select should involve environment by implication: it is not difficult to see that the term 'abstracting' implies 'abstracting from something' and so involves the environment as an implication.

The term 'abstractions of different orders' is, in this work, as fundamental as the term 'time-binding' was in the author's earlier *Manhood of Humanity*. Hence, it is impossible to be comprehensive about it at this stage; more will be forthcoming as we proceed.

But we have already come to some important semantic results. We have selected our structural metaphysics, and decided that in 1933 we should accept the metaphysics of 1933, which is given *exclusively* by science. We have decided to abandon the false to facts 'is' of identity and to use, instead, the best available language; namely, an actional, behaviouristic, functional, operational language, based on 'order'. And, finally, we have found a term which is functionally satisfactory and has the correct structural and neural implications, and which represents a *non-el* term, and of which the meanings can be expanded and refined indefinitely by assigning to them different orders.

In passing on to the general scientific outlook, similar structural remarks upon a *non-el* point of view apply, and are semantically of importance. Because of the *non-el* character of the work of the writers on the Einstein and new quantum theories, much use is made of this material in the present work. There is a marked structural, methodological, *and semantic* parallelism between all modern *non-el* strivings, which are extremely effective psycho-logically. More material on this subject is given in Parts IX and X.

Now, returning to the analysis of the object which we called 'pencil', we observe that, in spite of all 'similarities', this object is unique, is different from anything else, and has a *unique* relationship to the rest of the world. Hence, we should give the object a *unique name*. Fortunately, we have already become acquainted with the way mathematicians manufacture an endless array of individual names without unduly expanding the vocabulary. If we call the given object 'pencil$_1$', we could call another similar object 'pencil$_2$'. In this way, we produce individual names, and so cover the *differences*. By keeping the main root word 'pencil', we keep the implications of daily life, and also of *similarities*. The habitual use of such a device is structurally and semantically of extreme importance. It has already been emphasized repeatedly that our abstracting from physical objects or situations proceeds by missing, neglecting, or forgetting, and that those disregarded characteristics usually produce errors in evaluation, resulting in the disasters of life. If we acquire this extensional mathematical habit of using special names for unique individuals, we become conscious, not only of the similarities, but also of the differences, which consciousness is one of the mechanisms for helping the proper evaluation and so preventing or eliminating semantic disturbances.

So we now have before us a unique object which we call by a unique name 'pencil$_1$'. If we enquire what science 1933 has to say about this object, we find that this object represents structurally an extremely complex, dynamic process. For our purpose, which is *intuitive*, it is of little importance whether we accept the object as made up of atoms and the atom as made up of whirling electrons., or whether we accept the newer quantum theory, as given in Part X, according to which the atom is formulated in terms of 'electrons' but the 'electron' is the region where some waves reinforce each other, instead of being a 'bit' of something. It is of no importance from our point of view whether the atoms are of a finite size or whether they extend indefinitely and are noticeable to us only in the regions of reinforcement of the waves. Naturally, this last hypothesis has a strong semantic appeal, since it would account, when worked out, for many other facts, such as 'fulness', in a *non-el* language; but probably it would necessitate a postulation of some sub-electronic structures.

What is important for our *s.r* is that we realize the fact that the gross macroscopic materials with which we are familiar are *not* simply what we see, feel., but consist of dynamic processes of some extremely fine structure; and that we realize further that our 'senses' are not adapted to register these processes without the help of extra-neural means and higher order abstractions.

Let us recall, in this connection, the familiar example of a rotary fan, which is made up of separate radial blades, but which, when rotating with a certain velocity, gives the impression of a *solid disk*. In this case the 'disk' is not 'reality', but a nervous integration, or abstraction from the rotating blades. We not only see the 'disk' (*b*) where there is no disk, but, if the blades rotate fast enough, we could not throw sand through them, as the sand would be too slow to get through before being struck by one of the blades.

The 'disk' represents a *joint phenomenon* of the rotating blades (*a*) and of the abstracting power of our nervous system, which registers only the gross macroscopic aspects and slow velocities, but *not* the finer activities on subtler levels. We cannot blame 'the finite mind' for the failure to register the separate blades, because physical instruments may behave similarly. For instance, the illustrations (*a*) and (*b*) are photographs of a small fan which I use in lectures, and the photographic camera also missed the rotating blades and registered only a 'disk', in Fig. 1b.

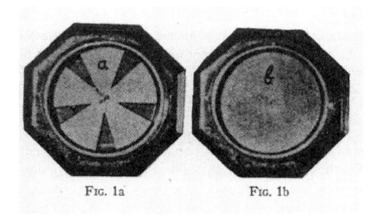

FIG. 1a   FIG. 1b

Something roughly similar may be assumed for our purpose as going on in what we usually call 'materials'. These are composed of some dynamic, fine-grained processes, not unlike the 'rotating blades' of our example; and what we register is the 'disk', be it a table or a chair or ourselves.

For a similar reason, we may assume that we cannot put our finger through a table, as our finger is too thick and too slow, and that, for some materials, it takes X-rays to be agile enough to penetrate.

The above analogies are helpful for our purpose only, but are oversimplified and should not be taken as a scientific explanation.

This neural process seems to be very general, and in all our daily experiences the dynamic fine structures are lost to our rough 'senses'. We register 'disks', although investigation discovers not 'disks', but rotating 'blades'. Our gross macroscopic experience is only a nervous abstraction of some definite order.

As we need to speak about such problems, we must select the best language at our disposal. This ought to be *non-el* and, structurally, the closest to facts. Such a language has been built, and is to be found in the differential and four-dimensional language of space-time, and in the new quantum mechanics. In practice, it is simple to ascribe to every 'point of space' a date, but it takes some training to get this *s.r.* The language of space-time is *non-el*. To the new notion of a 'point' in 'space-time', such a 'point', always having a date associated with it and hence never identical with any other point, the name of 'point-event', or simply 'event', has been given.

How to pass from point-events to extended macroscopic events is a problem in mathematical 'logic'. Several quite satisfactory schemes have been given, into the details of which we do not need to enter here. As the *non-el* structure of the language of space-time appears different from the older *el* language of 'space' *and* 'time', quite obviously the old term 'matter', which belonged to the descriptive apparatus of 'space' *and* 'time', should be abandoned also, and the 'bits' of materials we dealt with should be referred to by structurally new terms. In fact, we know that the old term 'matter' can be displaced by some other term connected with the 'curvature' of 'space-time'.

As abstracting in many orders seems to be a general process found in all forms of life, but particularly in humans, it is of importance to be clear on this subject and to select a language of proper structure. As we know already, we use *one* term, say 'apple', for at least *four* entirely different entities; namely, (1) the event, or scientific object, or the sub-microscopic physico-chemical processes, (2) the ordinary object manufactured from the event by our lower nervous centres, (3) the psychological picture probably manufactured by the higher centres, and (4) the verbal definition of the term. If we use a language of adjectives and subject-predicate forms pertaining to 'sense' impressions, we are using a language which deals with entities *inside our skin* and characteristics entirely non-existent in the outside world. Thus the events outside our skin are neither cold nor warm, green nor red, sweet nor bitter., but these characteristics are manufactured by our nervous system inside our skins, as responses only to different energy manifestations, physicochemical processes, . When we use such terms, we are dealing with characteristics which are absent in the external world, and build up an anthropomorphic and delusional world non-similar in structure to the world around us. Not so if we use a language of order, relations, or structure, which can be applied to sub-microscopic events, to objective levels, to semantic levels, and which can also be expressed in words. In using such language, we deal with characteristics found or discovered on all levels which give us *structural* data uniquely important for knowledge. The ordering on semantic levels in the meantime abolishes identification. It is of extreme importance to realize that the relational., attitude is optional and can be applied everywhere and always, once the above-mentioned benefits are realized. Thus, any object can be considered as a set of relations of its parts., any 'sense' perception may be considered as a response to a stimulus., which again introduces relations . As relations are found in the scientific sub-microscopic world, the objective world, and also in the psycho-logical and verbal worlds, it is beneficial to use such a language because it is *similar in structure* to the external world and our nervous system; and it is applicable to all levels. The use of such a language leads to the discovery of invariant relations usually called 'laws of nature', gives us structural data which make the only possible content of 'knowledge', and eliminates also anthropomorphic, primitive, and delusional speculations, identifications, and harmful *s.r.*

13. ON THE STRUCTURAL DIFFERENTIAL

If we take something, anything, let us say the object already referred to, called 'pencil', and enquire what it represents, according to science 1933, we find that the 'scientific object' represents an 'event', a mad dance of 'electrons', which is different every instant, which never repeats itself, which is known to consist of extremely complex dynamic processes of very fine structure, acted

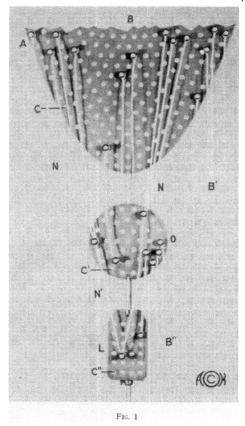

FIG. 1

upon by, and reacting upon, the rest of the universe, inextricably connected with everything else and dependent on everything else. If we enquire *how many characteristics* (m.o) we should ascribe to such an event, the only possible answer is that we should ascribe to an event infinite numbers of characteristics, as it represents a process which never stops in one form or another; neither, to the best of our knowledge, does it repeat itself.

In our diagram, Fig. 1, we indicate this by a parabola (A), which is supposed to extend indefinitely, which extension we indicate by a broken off line (B). We symbolize the characteristics by small circles (C), the number of which is obviously indefinitely great.

Underneath, we symbolize the 'object' by the circle (O), which has a finite size. The characteristics of the object we also denote by similar little circles (C'). The number of characteristics which an object has is large but *finite*, and is denoted by the finite number of the small circles (C').

Then we attach a label to the object, its name, let us say 'pencil₁', which we indicate in our diagram by the label (L). We ascribe, also, characteristics to the labels, and we indicate these characteristics by the little circles (C'').

The number of characteristics which we ascribe by *definition* to the label is still smaller than the number of characteristics the object has. To the label 'pencil₁' we would ascribe, perhaps, its length, thickness, shape, colour, hardness, . But we would mostly *disregard* the accidental characteristics, such as a scratch on its surface, or the kind of glue by which the two wooden parts of the objective 'pencil' are held together, . If we want an objective 'pencil' and come to a shop to purchase one, we say so and specify verbally only these characteristics which are of particular *immediate interest* to us.

It is clear that the object is often of interest to us for some special characteristics of immediate usefulness or value. If we enquire as to the neurological processes involved in registering the object, we find that the nervous system has *abstracted,* from the infinite numbers of sub-microscopic characteristics of the event, a large but finite number of macroscopic characteristics. In purchasing a 'pencil' we usually are not interested in its smell or taste. But if we were interested in these abstractions, we would have to find the smell and the taste of our object by experiment.

But this is not all. The object represents in this language a gross macroscopic abstraction, for our nervous system is not adapted for abstracting directly the infinite numbers of characteristics which the endlessly complex dynamic fine structure of the event represents. We must consider the object as a 'first abstraction' (with a finite number of characteristics) from the infinite numbers of characteristics an event has. The above considerations are in perfect accord not only with the functioning of the nervous system but also with its structure. Our nervous system registers objects with its lower centres first, and each of these lower specific abstractions we call an object. If we were to define an object, we should have to say that an object represents a first abstraction with a finite number of m.o characteristics from the infinite numbers of m.o characteristics an event has.

Obviously, if our inspection of the object is through the lower nervous centres, the number of characteristics which the object has is larger (taste, smell., of our pencil$_1$) than the number of characteristics which we need to ascribe to the label. The label, the *importance* of which lies in its *meanings to us*, represents a still higher abstraction from the event, and usually labels, also, a *semantic reaction*.

We have come to some quite obvious and most important structural conclusions of evaluation of the *non-el* type. We see that the object *is not* the event but an abstraction from it, and that the label *is not* the object nor the event, but a still further abstraction. The nervous process of abstracting we represent by the lines (N), (N'). The characteristics *left out*, or not abstracted, are indicated by the lines (B'), (B'').

For our semantic purpose, the distinction between lower and higher abstractions seems fundamental; but, of course, we could call the object simply the first order abstraction, and the label, with its meanings, the second order abstraction, as indicated in the diagram.

If we were to enquire how this problem of abstracting in different orders appears as a limiting case among animals, we should select a definite individual with which to carry on the analysis. For our analysis, which is deliberately of an extensional character, we select an animal with a definite, proper name, corresponding to 'Smith' among us. Such an animal suggests itself at once on purely verbal grounds.

It is the one we call 'Fido'. Practically all English speaking people are acquainted with the name 'Fido'. Besides, most of us like dogs and are aware of how 'intelligent' they are.

Investigations and experimenting have shown that the nervous system of a Fido presents, in structure and function, marked similarities to that of a Smith. Accordingly, we may assume that, in a general way, it functions similarly. We have already spoken of the event in terms of recognition; namely, that we can never recognize an event, as it changes continually. Whitehead points out the fundamental difference between an event and an object in terms of *recognition*; namely, that an event cannot be recognized, and that an object can be recognized. He defines the object as the recognizable part of the event. The use of this definition helps us to test whether Fido has 'objects'. Since experiments show that Fido can recognize, we have to ascribe to Fido objects by definition. If we enquire what the objects of Fido represent, the structure and function of his nervous system, which are very similar to ours, would suggest that Fido's objects represent, also, *abstractions* of some low order, from the events. Would his objects appear the 'same' as ours? No. First of all, the abstractions from events which we call objects are not the 'same', even when abstracted by different individuals among humans. An extreme example of this can be given in that limited form of colour-blindness which is called Daltonism, when an object which appears green to most persons appears red to the certain few who suffer from this disease. There is, at present, no doubt that the nervous abstractions of all organisms are individual, not only with each individual, but at different 'times' with one individual, and differ, also, for these higher groups (abstractions) which we call species. We can infer how the world appears to a particular organism only if its nervous structure is quite similar to our own. With species widely separated neurologically, such inferences are entirely unjustified. So, on general grounds, the 'objects' of Fido are not the 'same' as ours; on neurological grounds, they appear only similar. In daily experience, we know that we should have difficulty in recognizing our own glove among a thousand, but Fido could perform this detection for us much better. So the 'same' glove must have been registered in the nervous system of Fido differently from the way it has been in ours.

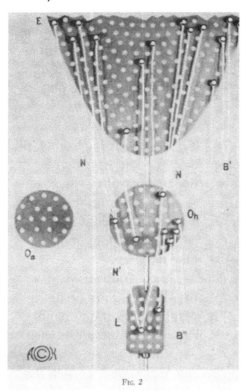

Fig. 2

We indicate this similarity of the human object (O_h) and the animal object (O_a) by making the circle (O_a) smaller, and emphasize the difference between the objects by differently spacing the holes representing the characteristics. Whether we call the objects (O_h) and (O_a) 'first order' abstractions or '100th order' abstractions, or simply 'lower order' abstractions, is mainly optional. There is no neurological doubt that all 'objects' represent *low order abstractions* and the use of a number to indicate the order is simply a matter of convention and convenience. If we were to start with the simplest living cell, we might ascribe to its abstractions the term 'first order' abstractions. If we were to survey in this way all known forms of life, we might ascribe to Fido and Smith very large numbers as their orders of abstractions. But this is unnecessary, as we shall presently see.

We note that Fido does abstract from events, at any rate, in lower orders, 'has objects' (O_a) which he can recognize. The question is, does he abstract in higher orders? We might answer that he does within certain limits. Or, we might prefer to take the limits of his abstracting capacities for granted and to include them all as lower order abstractions. For the sake of convenience and simplicity, we select the last method and say that he does not abstract in higher orders. In our schematic representation, we shall discover some very important differences between the abstracting capacities of humans and animals, and so we introduce here only as much complexity as we need. As animals have no speech, in the human sense, and as we have called the verbal labeling*of the object 'second order abstraction' we say that animals do not abstract in higher orders. (In the present system the terms 'label', 'labeling'., are always connected with their meanings, and so, for simplicity, from now on the reference to meanings will be omitted.)

If we compare our diagram and what it represents with the well-known facts of daily life, we see that Smith's abstracting capacities are not limited to two orders, or to any '*n*' orders of abstractions.

In our diagrams, the label (L) stands for the *name* which we assigned to the object. But we can also consider the level of the first label (L) as a *descriptive* level or statement. We know very well that Smith can always say something *about* a statement (L), on record. Neurologically considered, this *next* statement (L_1) about a statement (L) would be the nervous response to the former statement (L) which he has seen or heard or even produced by himself inside his skin. So his statement (L_1), *about* the former statement (L), is a *new abstraction* from the former abstraction. In my language, I call it an abstraction of a higher order. In this case, we shall be helped by the

*In the present system the terms 'label','labeling', are always connected with their meanings, and so, for simplicity, from now on the reference to meanings will be omitted.

use of numbers. If we call the level (L) an abstraction of *second order,* we must call an *abstraction from this abstraction* an abstraction of *third order,* (L_1). Once an abstraction of third order has been produced, it becomes, in turn, a fact on record, potentially a stimulus, and can be abstracted further and a statement made about it, which becomes an abstraction of the fourth order (L_2). This process has no definite limits, for, whenever statements of any order are made, we can always make a statement about them, and so produce an abstraction of still higher order. This capacity is practically universal among organisms which we call 'humans'. Here we reach a fundamental difference between 'Smith' and 'Fido'. Fido's *power of abstracting stops somewhere,* although it may include a few orders. Not so with 'Smith'; his power of abstracting has no known limit (see Part VI).

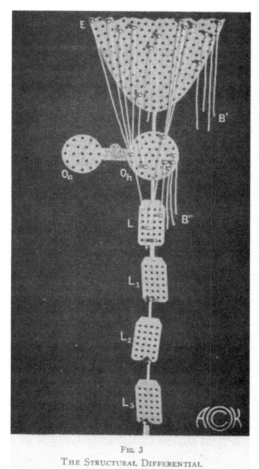

FIG. 3
THE STRUCTURAL DIFFERENTIAL

Perhaps the reader is semantically perplexed by the unfamiliarity of the language of this analysis. It must be granted that the introduction of any new language is generally perplexing, and it is justified *only* if the new language accomplishes something structurally and semantically which the old languages did *not* accomplish. In this case, it has brought us to a new *sharp* distinction between 'man' and 'animal'. The number of orders of abstractions an 'animal' can produce is *limited.* The number:' of orders of abstractions a 'man' can produce is, in principle, *unlimited.*

Here is found the fundamental mechanism of the 'time-binding' power which characterizes man, and which allows him, in principle, to gather the *experiences* of all past generations. A higher order abstraction, let us say, of the $n+1$ order, is made as a response to the stimulus of abstractions of the nth order. Among 'humans' the abstractions of high orders produced by others, as well as those produced by oneself are stimuli to abstracting in still higher orders. Thus, in principle, we start where the former generation left off. It should be noticed that, in the present analysis, we have abandoned the structurally *el* methods and language, and the whole analysis becomes simple, although non-familiar because it involves new *non-el s.r.*

The preceding explanation justifies my former statement that the ascribing of absolute numbers to the orders of abstractions of 'animal' and of 'man' is unnecessary. In our diagram we could ascribe as many orders of abstractions to the animal as we please; yet we should have to admit, for the structural correctness of description of experimental facts, that the 'animal's' power of abstracting has limits, while the number of orders of abstractions a 'man' can produce has no known limits.

From an epistemological and semantic point of view, there is an important benefit in this method. In this language, we have discovered *sharp* verbal and analytical methods, in terms of the *non-el* 'orders of abstractions', by which these two 'classes of life', or these two high abstractions, can be differentiated. The terms 'animal' and 'man' each represent a name for an abstraction of very high order, and not a name for an objective individual. To formulate the difference between these 'classes' becomes a problem of *verbal structural ingenuity and methods*, as in life we deal only with absolute individuals on the un-speakable, objective levels. In our diagram, we could hang on the 'animal' object as many levels of labels, which stand for higher order abstractions, as we please; yet somewhere we would have to stop; but with 'man' we could continue indefinitely.

This *sharp* difference between 'man' and 'animal' may be called the '*horizontal difference*'. The habitual use of *our hands* in showing these different horizontal levels is extremely useful in studying this work, and it facilitates greatly the acquiring of the structurally new language and corresponding *s.r.* The solution of the majority of human semantic difficulties (evaluation), and the elimination of pathological identification, lie precisely in the maintenance, without confusion, of the sharp differentiation between these horizontal levels of orders of abstractions.

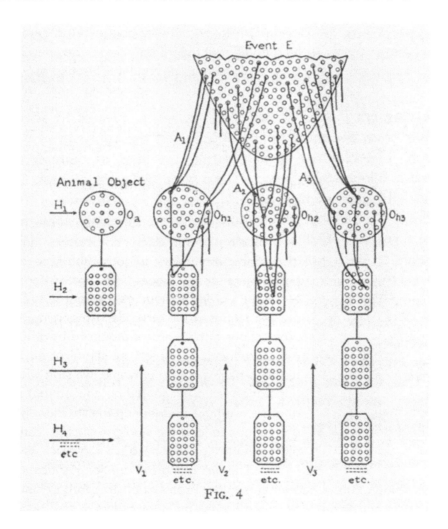

Fig. 4

Let us now investigate the possibility of a sharp '*vertical difference*'. We have already come to the conclusion that Fido abstracts objects from events, and that, if his nervous system is similar to ours, his lower order abstractions are similar to ours. Here we may ask the question: Does Fido 'know', or can he 'know', that he abstracts? It seems undeniable that Fido does not 'know' and *cannot 'know'* that he abstracts, *because it takes science to 'know'* that we abstract, and Fido has no science. It is semantically important that we should be entirely convinced on this point. We do not argue about the kind of 'knowledge' animals may have or about the relative value of this 'knowledge' as compared with ours. Science was made possible by the human nervous system and the invention of extra-neural means for investigation and recording, which animals lack entirely. Whoever claims that animals have science should, to say the least, show libraries and scientific laboratories and instruments produced by animals.

We see that, although Fido has abstracted, he not only does not 'know' but *cannot* 'know' that he abstracts, as this last 'knowledge' is given exclusively by science, which animals do not have. *In this consciousness of abstracting, we find a most important 'vertical difference' between Smith and Fido.* The difference is sharp again.

If, in our diagram, Fig. 4, we ascribe to Fido more horizontal orders of abstractions, let us say two, (H_1) and (H_2), nevertheless, the 'animal' stops somewhere. This extended diagram illustrates that 'man' is capable of abstracting in higher and higher orders indefinitely. In this diagram, we symbolize the fact that Fido does not and cannot 'know' that he abstracts, by not connecting the characteristics of his object (O_a) by lines (A_n) with the event (E). *Without science, we have no event; Fido's gross macroscopic object (O_a) represents 'all' that he 'knows' or cares about.* We see that the *vertical difference* (V_1) formulated as consciousness of abstracting for Smith appears sharp, and completely differentiates Fido from Smith. In it, we find the semantic mechanism of all proper *evaluation*, based on *non-identification* or the differentiation between orders of abstractions, impossible with animals.

In this diagram we have introduced more objects, because each individual abstracts, in general, from an event *different* objects, in the sense that they are *not identical* in every respect. We must be aware continuously that in life on the un-speakable objective level we deal only with absolute individuals, be they objects, situations, or s.r. The vertical stratification not only gives us representation for the sharp difference between 'man' and 'animal', but also allows us to train our s.r in the absolute individuality of our objects and those of different observers, and for the differences between their individual abstractions. What has been said here applies equally to all first order effects on the objective level, such as immediate feelings, .

The present theory can only be fully beneficial when the reader acquires in *his system* the habitual feeling of both the vertical and the horizontal stratifications with which identification becomes impossible. In the experiments of Doctor Philip S. Graven with the 'mentally' ill, training in the realization of this stratification has either resulted in complete recovery or has markedly improved the conditions of the patient.

The diagram is used in *two* distinct ways. One is by showing the abstracting from the event to the object, and the applying of a name to the object. The other is by illustrating the level of statements which can be made about statements. If we have different objects, and label them with different names, say, A_1, A_2, A_3 . . . A_n, . we still have no proposition. To make a proposition, we have to accept some undefined relational term, by which we relate one object to the other. The use of this diagram to illustrate the *levels or orders of statements* implies that we have selected some metaphysics as expressed in our undefined relational terms. We should be fully aware of the difference between these *two* uses of the one diagram for the structural illustration of two aspects of one process.

If we enquire: What do the characteristics of the event represent? We find that they are given only by science and represent at each date the highest, most verified, most reliable abstractions 'Smith' has produced.

Theory and practice have shown that the points illustrated by the above structural diagrams have a crucial semantic significance, as, without using them, it is practically impossible to train ourselves or others and to accomplish the psychophysiological re-education. For this reason, the diagrams have been produced for home and school use, separately, in the simplified form illustrated in Fig. 5. This structural diagram is called the 'Anthropometer' or the 'Structural Differential', as it illustrates the fundamental structural difference between the world, *and so the environment*, of the animal and man. If we live in such a very complex *human* world, but our *s.r*, owing to wrong *evaluation*, are adjusted only to the simpler animal world, free, to say the least, from man-made complications, then adjustment and sanity for *humans* is impossible. Our *s.r* are bound to follow the simpler animalistic patterns, *pathological for man*. All human experience, scientific or otherwise, shows that we still copy animals in our nervous reactions, trying to adjust ourselves to a world of fictitious, simple *animal structure*, while *actually* we live in a world of very complex *human structure* which is quite different. Naturally, under such conditions, which, ultimately, turn out to be delusional, *human* adjustment is impossible and results in false evaluations, animalistic *s.r*, and the general state of un-sanity.

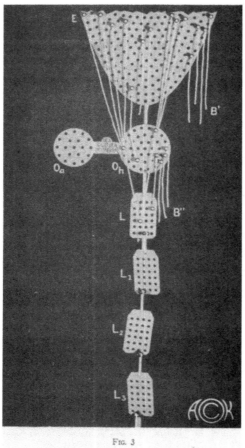

FIG. 3

THE STRUCTURAL DIFFERENTIAL

Any one who will work out the present analysis with the aid of the Differential will find clearly that the majority of human difficulties, the preventable or curable 'mental' or semantic disturbances included, are due to this fatal *structural* error, resulting in false evaluation due to identification or lack of differentiation.

For the event we have a parabola in relief (E), broken off to indicate its limitless extension. The disk (O_h) symbolizes the human object; the disk (O_a) represents the animal object. The label (L) represents the higher abstraction called a name (with its meaning given by a definition). The lines (A_n) in the relief diagram are hanging strings which are tied to pegs. They indicate the process of abstracting. The free hanging strings (B_n) indicate the most important characteristics *left out*, neglected, or forgotten in the abstracting. The Structural Differentials are provided with a number of separate labels attached to pegs. These are hung, one to the other, in a series, and the last one may be attached by a long peg to the event, to indicate that the characteristics of the event represent the highest abstractions we have produced at each date. *The objective level is not words, and cannot be reached by words alone. We must point our finger and be silent, or we shall never reach this level.* Our personal feelings, also, *are not* words, and belong to the objective level.

The whole of the present theory can be illustrated on the Structural Differential by the childishly simple operation of the teacher pointing a finger to the event and then to the object, saying 'This *is not* this' and insisting on silence on the pupil's part. One should continue by showing with the finger the object and the label, saying again 'This *is not* this', *insisting on silence* on the objective level; then, showing the first and the second label, saying again 'This *is not* this', .

In a more complex language, one would say that the object *is not* the event, that the label *is not* the un-speakable object, and that a statement about a statement *is not* the 'same' statement, nor on one level. We see and are made to visualize that the [non-A]-system is based *on the denial of the 'is' of identity*, which necessitates the differentiation of orders of abstractions.

The little word 'to be' appears as a very peculiar word and is, perhaps, responsible for many human semantic difficulties. If the anthropologists are correct, only a few of the primitive peoples have this verb. The majority do not have it and do not need it, because all their *s.r* and languages are practically based on, and involve, literal *identification*. In passing from the primitive stage of human society to the present slightly higher stage, which might be called the infantile stage, or infantile period, too crude identification was no longer possible. Languages were built, based on slightly modified or limited identification, and, for flexibility, the 'is' of identity was introduced explicitly. Although very little has been done in the *structural* analysis of languages in general, and of those of primitive peoples in particular, we know that in the Indo-European languages the verb 'to be', among others, is used as an *auxiliary verb* and also for the purpose of positing false to facts identity. With the primitive prevalent lack of consciousness of abstracting, and the primitive belief in the magic of words, the *s.r* were such that words were identified with the objective levels. Perhaps it is not too much to say that the primitive 'psychology' peculiarly required such a fundamental identity. Identity may be defined as 'absolute sameness in all respects' which, in a world of ever-changing processes and a human world of indefinitely many orders of abstractions, appears as a *structural* impossibility. Identity appears, then, as a primitive 'over-emotional' generalization of similarity, equality, equivalence, equipollence., and, in no case, does it appear in fact as 'absolute sameness in all respects'. As soon as the structurally *delusional* character of identity is pointed out, it becomes imperative for sanity to eliminate such delusional factors from our languages and *s.r.* With the advent of 'civilization', the use of this word was enlarged, but some of the fundamental primitive implications and psycho-logical semantic effects were preserved. If we use the 'is' at all, and it is extremely difficult to avoid entirely this auxiliary verb when using languages which, to a large extent, depend on it, we must be particularly careful not to use 'is' as an identity term.

In 1933, it seems, beyond doubt, that *if* any single *semantic* characteristic could be selected to account for the primitive state of the individuals and their societies, we could say, without making too great a mistake, that it would be found in *identification*, understood in the more general sense as it is used in the present work. There is very little doubt, at present, that different physico-chemical factors, environment, climate, kind of food, colloidal behaviour, endocrine secretions., are fundamental factors which condition the potentialities, as well as the behaviour, of an organism. It is equally certain that, as an end-result, these physico-chemical factors are connected with definite types of *s.r.* It is known that the reverse is also true; namely, that *s.r* affect colloidal behaviour, endocrine secretions, and metabolism. The exact type of dependence is not known, because too little experimenting on humans has been made. The present analysis is conducted from the semantic point of view, and its results, no matter how far-reaching, are limited to this special aspect.

Simple analysis shows that identification is a necessary condition which underlies the reactions of animals, of infants, and of primitives. If found in 'civilized' grown-ups, it equally indicates some remains of earlier periods of development, and can always be found in the analysis of any private or public difficulties which prevent any satisfactory solution. Identification in a slightly modified form represents, also, the very foundation of the [A]-system and those institutions which are founded on this system.

Mathematics gives us practically the only linguistic system free from pathological identifications, although mathematicians use this term uncritically. The more identification is eliminated from other sciences, the more the mathematical functional semantics and method are applied, and the further a given science progresses.

The best we know in 1933 is that the general structure of the world was not different in prehistoric times from what we find it today. We have no doubt that the materials in great antiquity consisted of molecules, molecules of atoms, and atoms of electrons and protons., or whatever else we shall be able to discover some day. We have no doubt that blood was circulating in the higher animals and humans, that vitamins exhibited very similar characteristics as today, that different forms of radiant energy influenced colloidal behaviour., ., regardless of whether or not the given animal, primitive man or infant 'knew' or 'knows' about them.

How about the primitive *physical needs and wants* of an animal, a primitive man, and an infant? Besides all mystical and mythological reasons for identification, the structural facts of life *necessitated identification* on this level of development. *Without* modern knowledge, what a hungry animal, primitive man, or an infant 'wants' 'is' an 'object', say, called an 'apple'. He would 'define' his 'apple' the best he could as to shape, colour, smell, taste, . Was this what his organism needed? Obviously not. We could, at present, produce an undigestible synthetic apple which would satisfy his eventual objective definitions; he might eat it, many such 'apples', and eventually die of hunger. Is an abundant and pleasant diet free from unsuspected and unseen 'vitamins' satisfactory for survival? Again, no ! Thus, we see clearly that what the organism needed for survival were the physico-chemical processes, not found in the 'ordinary object', but exclusively in the 'scientific object', or the event. Here we find the age old and necessary, on this early level, identification of the ordinary object with the scientific object. This form of identification is extremely common even in 1933, and, to a large extent, responsible for our low development, because, no matter what we 'think' or feel about an object, an object represents *only* an abstraction of low order, only a *general symbol* for the *scientific object,* which remains the only possible survival concern of the organism. But, obviously, such identification, being false to facts, can never be entirely reliable. If any one fancies that he deals with 'ultimate reality', yet that *m.o* reality represents only a shadow cast by the scientific object; he begins, with experience, to distrust the object and populates his world with delusional mysticism and mythologies to account for the mysteries of the shadow.

As any organism represents an *abstracting* in *different orders* process, which, again, the animal, the primitive man, and the infant cannot know, they, by necessity, identify different orders of abstractions. Thus, names are identified with the un-speakable objects, names for action with the un-speakable action itself, names for a feeling with the unspeakable feelings themselves. By confusing descriptions with inferences and descriptive words with inferential words, the 'judgements', 'opinions', 'beliefs', and similar *s.r,* which represent mostly, if not exclusively, inferential semantic end-products, are projected with varying pathological intensity on the outside world. By this method pre-'logical' primitive semantic attitudes were built. Mere similarities were evaluated as identities, primitive syllogisms were built of the type: 'stags run fast, some Indians run fast, some Indians are stags'. It is common to find among primitive peoples a kind of 'logic' based on the *post hoc, ergo propter hoc* (after this and, therefore, because of this) fallacy which obviously represents an identification of an ordinal description with an inference. The 'question begging epithets', which exercise a tremendous semantic influence on primitive and immature peoples and represent a semantic factor in many primitive as well as modern taboos, are also based on such confusions of orders of abstractions.

Identification is one of the primitive characteristics which cannot be eliminated from the animal or the infant, because we have no means to communicate with them properly. It cannot be eliminated from primitive peoples as long as they preserve their languages and environments. Identification is extremely wide-spread among ourselves, embodied strongly in the structure of our inherited language and systems. To change that primitive state of affairs, we need special simple means, such as a [non-A]-system may offer, to combat effectively this serious menace to our *s.r.* It should never be forgotten that identification is practically never dangerous in the animal world, because unaided nature plays no tricks on animals and the elimination by non-survival is very sharp. It is dangerous in the primitive stage of man, however, as it prevents the primitive man to become more civilized, but under his primitive conditions of life his dangers are not so acute. It becomes only very dangerous to the infant if not taken care of and to the modern white man in the midst of a very far advanced industrial system which affects all phases of his life, when his *s.r* are left unchanged from the ages gone by, and still remain on the infantile level.

The present [non-A]-system is not only based on the complete rejection of the 'is' of identity, but every important term which has been introduced here, as well as the Structural Differential, is aimed at the elimination of these relics of the animal, the primitive man, and the infant in us.

Thus, the primitive 'mentality' does *not differentiate relations* enough; to counteract this, I introduce the *Structural Differential*. The primitive identifies; I introduce a system based on the denial of the 'is' of identity all through. The primitive man pays most attention to what is conveyed to him through the eye and the ear; I introduce the Structural Differential which indicates to the eye the stratification of human knowledge, which represents to the eye the verbal denial of the 'is' of identity. If we identify, we do not differentiate. If we differentiate, we cannot identify; hence, the Structural Differential.

The terms used also convey similar processes. Once we have *order*, we differentiate and have orders of abstractions. Once we abstract, we eliminate 'allness', the semantic foundation for identification. Once we abstract, we abstract in different orders, and so we *order*, abolishing fanciful infinities. Once we differentiate, differentiation becomes the denial of identity. Once we discriminate between the objective and verbal levels, we learn 'silence' on the un-speakable objective levels, and so introduce a most beneficial neurological 'delay'—engage the cortex to perform its natural function. Once we discriminate between the objective and verbal levels, structure becomes the only link between the two worlds. This results in search for similarity of structure and relations, which introduces the aggregate feeling, and the individual becomes a *social being*. Once we differentiate, we discriminate between descriptions and inferences. Once we discriminate, we consider descriptions separately and so are led to *observe* the facts, and only from description of facts do we tentatively form inferences, . Finally, the consciousness of abstracting introduces the general and permanent differentiation between orders of abstractions, introduces the ordering, and so stratifications, and abolishes for good the primitive or infantile identifications. The semantic passing from the primitive man or infantile state to the adult period becomes a semantic, accomplished fact. It should be noticed that these results are accomplished by starting with primitive means, the use of the simplest terms, such as 'this *is not* this', and by the direct appeal to the primitive main receptors—the eye and the ear.

The elimination of the 'is' of identity appears as a serious task, because the [A]-system and 'logic' by which we regulate our lives, and the influence of which has been eliminated only partially from science, represent only a very scholarly formulation of the restricted primitive identification. Thus, we usually assume, following [A] disciplines, that the 'is' of identity is fundamental for the 'laws of thought', which have been formulated as follows:

1) The Law of Identity: whatever is, is.

2) The Law of Contradiction: nothing can both be and not be.

3) The Law of Excluded Middle: everything must either be or not be.

It is impossible, short of a volume, to revise this 'logic' and to formulate a [non-A], ∞-valued, *non-elementalistic* semantics which would be structurally similar to the world and our nervous system; but it must be mentioned, even here, that the 'law of identity' is never applicable to processes. The 'law of excluded middle', or 'excluded third', as it is sometimes called, which gives the two-valued character to [A] 'logic', establishes, as a general principle, what represents only a limiting case and so, *as a general principle*, must be unsatisfactory. As on the objective, un-speakable levels, we deal exclusively with absolute individuals and individual situations, in the sense that they are not identical, all statements which, by necessity, represent higher order abstractions must only represent *probable* statements. Thus, we are led to ∞-valued semantics of probability, which introduces an inherent and general principle of uncertainty.

It is true that the above given 'laws of thought' can and have been expressed in other terms with many scholarly interpretations, but fundamentally the semantic state of affairs has not been altered.

From a *non-el* point of view, it is more expedient to treat the [A]-system on a similar footing with the [E]-system; namely, to consider the above 'laws of thought' as postulates which underlie that system and which express the 'laws of thought' of a given epoch and, eventually, of a race. We know other systems among the primitive peoples which follow other 'laws', in which identity plays a still more integral part of the system. Such natives reason quite well; their systems are consistent with their postulates, although these are quite incomprehensible to those who try to apply [A] postulates to them. From this point of view, we should not discuss how 'true' or 'false' the [A]-system appears, but we should simply say that, at a different epoch, other postulates seem structurally closer to our experience and appear more expedient. Such an attitude would not retard so greatly the appearance of new systems which will supersede the present [non-A]-system.

In the present system, 'identification' represents a label for the semantic process of inappropriate evaluation on the un-speakable levels, or for such 'feelings', 'impulses', 'tendencies',. As in human life, we deal with many orders of abstractions, we could say in an ordinal language that identification originates *or* results in the confusion of orders of abstractions. This conclusion may assume different forms: one represented by the identification of the scientific object or the event with the ordinary object, which may be called ignorance, pathological to *man*; another, the identification of the objective levels with the verbal levels, which I call objectification; a third, the identification of descriptions with inferences, which I call confusion of higher order abstractions. In the latter case, we should notice that inferences involve usually more intense semantic components, such as 'opinions', 'beliefs', 'wishes'., than descriptions. These inferences may have a definite, objective, un-speakable character and may represent, then, a semantic state which *is not* words, and so objectifications of higher order may be produced.

When we introduce the ordinal language, we should notice that under known conditions we deal with an ordered natural series; namely, events first, object next; object first, label next; description first, inferences next, . This order expresses the natural importance, giving us the natural base for evaluation and so for our natural *human s.r.* If we identify two different orders, by necessity, we evaluate them equally, which always involves errors, resulting potentially in semantic shocks. As we deal in life with an established natural order of values which can be expressed, for my purpose, by a series decreasing in value: events or scientific objects, ordinary objects, labels, descriptions, inferences.; identification results in a very curious semantic situation.

Let us assume that the scientifically established value of any level could be expressed as

100, and the value of the next as 1. With the consciousness of abstracting we could not disregard, nor identify, these values, nor forget that 100>1. If we confuse the orders of abstractions, this can be expressed as the identification in value and we have a *semantic* equation: (1) 100=100, or (2) 1=1, or any other number, say (3) 50=50.

As we deal fundamentally with a natural, directed inequality, say, 100>1, and, under some semantic pressure, 'want', 'wishful thinking', or ignorance, or lack of consciousness of abstracting, or 'mental' illness., we identified the two in value, we produce in the first and third cases an over-evaluation on the right-hand side, and, in the second and third cases, an *under*-evaluation on the left-hand side. Thus, on the *semantic level*, any identification of *essentially different in value* different orders of abstractions, appears as the *reversal of* the natural order of evaluation, with different degrees of intensity. If the *natural* order of scientific evaluation would be 100>1, and we would evaluate through identification as 2=2, or 3=3., 50=50., 100=100, we would be ascribing twice, or three times, or fifty times, or a hundred times., more *delusional* values to the right-hand side and under-evaluate the left-hand side, than the natural order of evaluation would require. Nature exhibits, in my language and in this field, an asymmetrical relation of 'more', or 'less' inaccessible to [A] procedure. Under the influence of aristotelianism, when, through identification, we ascribe to nature delusional values, adjustment becomes very difficult, particularly under modern complex life-conditions.

The above example indicates the degrees of intensity which we find in life in the reversal of the natural order of evaluation through identification, produced by, and resulting in, the lack of consciousness of abstracting. Un-sanity, which affects practically all of us, represents the reversal of lesser intensity; the reversal of greater intensity—the more advanced 'mental' ills.

We should realize that *experimentally* we find in this field a fundamental difference in value, which, on semantic levels, can be expressed as an asymmetrical relation of 'more' or 'less', establishing some natural order. If any one should claim a natural 'identity', the burden of proof falls on him. If 'absolute sameness in all respects' cannot be found in this world, then such a notion appears as false to facts, and becomes a structural falsification, preventing sanity and adjustment. If he accepts the fundamental, natural differences in value, but prefers to assume a different order of evaluation depending on his metaphysics, be it the *elementalistic* materialism, or equally *elementalistic* idealism, the semantic results are not changed, because identification in the second case would also ascribe delusional identity to essentially different orders of abstractions. It should be noticed that the [non-A] formulation applies equally to the older different, opposite doctrines and renders them illegitimate on similar grounds.

The status of the event, or the scientific object, is slightly more complex, because the event is *described* at each date by very reliable, constantly revised and tested, *hypothetical*, structural, inferential terms, exhibiting the peculiar circularity of human knowledge. If we should treat these inferential structures, not as hypothetical, but should identify them semantically with the eventual processes on the level of the submicroscopic event, we would have semantic disturbances of identification.

I have selected the above given order, not only for convenience and simplicity, but because of its experimental character. When we identify in values, we always exhibit in our *s.r* the reversed natural order, introduced here on space-time structural and evaluational grounds.

14. ON 'CONSCIOUSNESS' AND CONSCIOUSNESS OF ABSTRACTING

We have already discovered *functional* differences that are expressed by the horizontal and by the vertical differences between the abstracting capacities of Smith and Fido. The analysis of these differences is the subject of the present chapter.

'Thought' represents a reaction of the organism-as-a-whole, produced by the working of the whole, and influencing the whole. From our daily experience, we are familiar with what we usually denote as being 'conscious'; in other words, we are aware of something, be it an object, a process, an action, a 'feeling', or an 'idea'. A reaction that is very habitual and semi-automatic is not necessarily 'conscious'. The term 'consciousness', taken separately, is not a complete symbol; it lacks content, and one of the characteristics of 'consciousness' is to have some content. Usually, the term 'consciousness' is taken as undefined and *undefinable*, because of its immediate character for every one of us. Such a situation is not desirable, as it is always semantically useful to try to define a complex term by simpler terms. We may limit the general and undefined term 'consciousness' and make it a definite symbol by the deliberate ascribing of some content to this term. For this 'consciousness of something' I take 'consciousness of abstracting' as fundamental. Perhaps the only type of meanings the term 'consciousness' has is covered by the functional term 'consciousness of abstracting', which represents a general process going on in our nervous system. Even if this is not the only type of meanings, the term 'consciousness of abstracting' appears to be of such crucial semantic importance that its introduction is necessary.

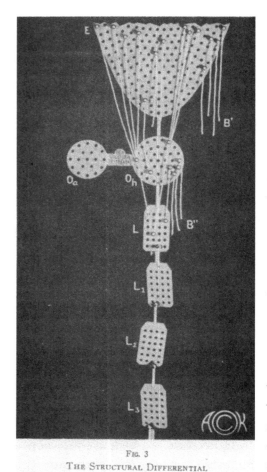

Fig. 3

The Structural Differential

The term 'consciousness', because of its hitherto undefined and traditionally *undefinable* character, did not allow us further analysis. Neither did we have any *workable*, educational, semantic means to handle the vast field of psycho-logical processes which this incomplete symbol indicated. If we now select the term 'consciousness of abstracting' as fundamental, we not only make the last symbol complete by assigning functional content to it, but we also find means to define it more specifically in *simpler terms*. Through understanding of the processes we gain educational means of handling and influencing a large group of semantic psycho-logical reactions.

Let us analyse this new term by aid of the diagram called the Structural Differential referred to in the previous chapter. Here the object (O_h) represents a nervous abstraction of a low order. In this abstracting, some characteristics of the event were missed or not abstracted; these are indicated by the not connected lines (B'). When we abstracted from our object further, by coining a definition or ascribing 'meanings' to the label (L), again we did not abstract 'all' the characteristics of the object into the definition; but some characteristics were left out, as indicated by the lines (B''). In other words, the number of characteristics which we ascribe to the label, by some process of 'knowing', or 'wanting', or 'needing', or 'interest'., does *not* cover the number of characteristics the object has. The 'object' has more characteristics than we can include in the explicit or implicit definition of the label for the 'object'. Besides, the definition (implicit or explicit) of the 'object' *is not* the object itself, which always holds many surprises for us. The latter has the *individuality of the object'*, as we may call it. Every one who uses a car, or a gun, or a typewriter, or who has had a number of wives, or husbands, or children, knows that well. In spite of the fact that these objects are, to a large extent, standardized, every individual object has individual peculiarities. With modern methods of physical, chemical, and astronomical investigation, scientists find that even their special materials and equipments have also peculiar individualities which must be taken into account in the more refined researches.

If we take any ordinary object and expect to find such and such characteristics, ascribed to the objects by *definition,* we may be disappointed. As a rule, we find or can find, if our analysis is subtle enough, these peculiar individualities. The reader can easily convince himself by looking over a box of matches, and by noticing the peculiar individuality of each match. But since, *by definition,* we expect that when we strike a match it should ignite, we may disregard all other characteristics as irrelevant for our purpose. A similar process is at work in other phases of life. We often live, feel happy or unhappy, *by what actually amounts to a definition,* and not by the empirical, individual facts less coloured by semantic factors. When Smith$_1$ marries Smith$_2$, they mostly do so *by a kind of definition*. They have certain notions as to what 'man', 'woman', and 'marriage' 'are' *by definition.* They actually go through the performance and find that the Smith$_1$ and his wife, Smith$_2$, have unexpected likes, dislikes, and particularities—in general, characteristic and semantic reactions *not included* in their definition of the terms 'man', 'woman', 'husband', 'wife', or 'marriage'. Characteristics 'left out' in the definitions make their appearance. 'Disappointments' accumulate, and a more or less unhappy life begins.

The above analysis applies to all phases of human life, and appears entirely general because of the structure of 'human knowledge'. Characteristics are discovered when it is *too late.* The *not knowing* or the *forgetting* of the relations explained above does the semantic havoc. On verbal, 'definitional', or doctrinal semantic grounds, we expect something else than what the experiences of life give us. The non-fulfillment of expectation produces a serious affective and semantic shock. If such shocks are repeated again and again, they disorganize the normal working of the nervous system, and often lead to pathological states. An indefinitely large number of experimental facts fully supports the above conclusions. Many of them have been supplied during the World War. Curiously enough, when the soldier *did* expect horrors, and later experienced them, he seldom became deranged 'mentally'. If he did not fully expect them, and yet had to experience them, he often broke down nervously.

The attack of hay fever at the sight of *paper roses,* referred to already, gives a similar semantic example. The attack followed from the semantic *'definition'* of 'roses', of 'hay fever', and from the situation as-a-whole, and was not due to *inspection* of the objective 'roses', or to the physico-chemical action of the 'roses'. If the patient had been blindfolded when the paper 'roses' were brought into his presence, no attack would have occurred.

We are now ready to define 'consciousness of abstracting' in *simpler terms;* namely, in terms of 'memory'. The term 'memory' is structurally a physico-chemical term. It implies that the events are interconnected, that everything in this world influences everything else, and that happenings leave some traces somewhere.

A similar analysis can be carried on in connection with the object and the event. Briefly, the object represents structurally an abstraction of some order, does not, and cannot, include all the characteristics of the event; and so, again, we have some characteristics *left out* as indicated by the lines (B').

Here we have the possibility of making a series of most general, and yet entirely true, *negative* statements of great semantic importance; that the label *is not* the object, and that the object *is not* the event, . For the number of m.o characteristics which we ascribe to the label by *definition* does not cover all the characteristics we recognize in the object; and the number of characteristics which we perceive in the object is also not equal to the infinite numbers of characteristics the event has. The differences are still more profound. Not only do the numbers of m.o characteristics differ, but also the *character* of these abstractions differs from level to level of the successive abstractions.

We can now define 'consciousness of abstracting' as '*awareness* that in our process of abstracting we have *left out* characteristics'. Or, consciousness of abstracting can be defined as '*remembering* the "*is not*", and that some characteristics have been *left out*'. It should be noticed that in this formulation, with the aid of the Structural Differential, we have succeeded in translating a *negative* process of forgetting into a *positive* process of *remembering* the denial of identity and that characteristics are left out. Such a positive formulation makes the whole system workable and available for the semantic training and education.

The use of the Structural Differential becomes a necessity for any one who wants to receive full semantic benefit from the present work. A book is, by necessity, *verbal*. Whatever any author can say is verbal, and nothing whatsoever can be *said* which is *not verbal*. It seems entirely obvious that in life we deal with an enormous number of things and situations, 'feelings'., which are *not verbal*. These belong to the 'objective level'. The crucial difficulty is found in the fact that whatever can be said *is not* and *cannot* be on the objective level, but belongs *only* to the verbal levels. This difference, being *inexpressible* by words, cannot be expressed by words. We must have *other means* to indicate this difference. We must show with our hand, by pointing our finger to the object, and by being silent outwardly as well as inwardly, which silence we may indicate by closing our lips with the other hand. The verbal denial of the 'is' of identity covers this point also when shown on the Differential. If we burst into speech based on the 'is' of identity, as we usually do, we find ourselves obviously on the verbal levels indicated by the labels L, L_1, L_2, . . . L_n., but never on the objective level (O_h). On this last level, we can look, handle., but *must be silent*. The reason that we nearly all identify the two levels is that it is impossible to train an individual in this semantic difference by *verbal means alone*, as all verbal means belong to the levels of labels and never to the objective unspeakable levels. With a visual and tactile *actual object* and labels on the Structural Differential, to point our finger at, handle., we now have simple means to convey the tremendously important semantic difference and train in *non-identity*.

We should notice that the consciousness of abstracting, or the remembering that we abstract in different orders with omission of characteristics, depends on the denial of the 'is' of identity and is connected with limitations or 'non-allness', so characteristic of the new non-systems.

The consciousness of abstracting eliminates *automatically* identification or 'confusion of the orders of abstractions', both applying to the semantic confusion on all levels. If we are *not* conscious of abstracting, we are bound to identify or confuse the object with its finite number of characteristics, with the event, with its infinite numbers of *different* characteristics. Confusion of these levels may misguide us into semantic situations ending in unpleasant shocks. If we acquire the consciousness of abstracting, and remember that the object *is not* the event and that we have abstracted characteristics fewer than, and different from, those the event has, we should expect many unforeseen happenings to occur. Consequently, when the unexpected happens, we are saved from painful and harmful semantic shocks.

If, through lack of consciousness of abstracting, we identify or confuse words with objects and feelings, or memories and 'ideas' with experiences which belong to the un-speakable objective level, we identify higher order abstractions with lower. Since this special type of semantic identification or confusion is extremely general, it deserves a special name. I call it *objectification*, because it is generally the confusion of words or verbal issues (memories, 'ideas'.,) with objective, un-speakable levels, such as objects, or experiences, or feelings, . If we objectify, we *forget*, or we *do not remember* that words *are not* the objects or feelings themselves, that the verbal levels are always different from the objective levels. When we identify them, we disregard the inherent differences, and so proper evaluation and full adjustment become impossible.

Similar semantic difficulties arise from the confusion of higher order abstractions; for instance, the identification of inferences with descriptions. This may be made clearer by examples. In studying these examples, it should be remembered that the organism acts as-a-whole, and that 'emotional' factors are, therefore, always present and should not be disregarded. In this study, the reader should try to put himself *'emotionally'* in the place of the Smith we speak about; then he cannot fail to understand the serious semantic disturbances these identifications create in everybody's life.

Let us begin with a Smith who knows nothing of what has been said here, and who is *not* conscious of abstracting. For him, as well as for Fido, there is, in principle, no realization of the 'characteristics left out'. He is 'emotionally' convinced that his words entirely cover the 'object' which '*is* so and so'. He identifies his lower abstractions with characteristics left out, with higher abstractions which have all characteristics included. He ascribes to words an entirely false value and certitude which they cannot have. He does not realize that his words may have different meanings for the other fellow. He ascribes to words 'emotional' *objectivity* and value, and the verbal, [A] 'permanence', 'definiteness', 'one-value'., to objects. When he hears something that he does not like, he does not ask 'what do you mean?', but, under the semantic pressure of identification, he ascribes his own meanings to the other fellow's words. For him, words '*are*' 'emotionally' overloaded, objectified semantic fetishes, even as to the primitive man who believed in the 'magic of words'. Upon hearing anything strange, his s.r is undelayed and may appear as, 'I disagree with you', or 'I don't believe you', . There is no reason to be dramatic about any unwelcome statement. One needs definitions and interpretations of such statements, which probably are correct from the speaker's point of view, if we grant him his informations, his *undefined terms*, the structure of his language and premises which build up his s.r. But our Smith, innocent of the 'structure of human knowledge', has mostly a semantic belief in the one-value, absoluteness., of things, and thinghood of words, and does not know, or does not *remember*, that words *are not* the events themselves. Words represent higher order abstractions manufactured by higher nerve centres, and objects represent lower order abstractions manufactured by lower nerve centres. Under such *identity-delusions*, he becomes an absolutist, a dogmatist, a finalist, . He seeks to establish 'ultimate truths', 'eternal verities'., and is willing to fight for them, never knowing or remembering, otherwise forgetting, the 'characteristics left out'; never recognizing that the noises he makes *are not* the objective actualities we deal with. If somebody contradicts him, he is much disturbed. Forgetting characteristics left out, he is always 'right'. For him his statement is not only *the* only statement possible, but he actually attributes some cosmic objective evaluation to it.

The above *description* is unsatisfactory, but cannot be much improved upon, since the situation involves *un-speakable* affective components which *are not* words. We must simply try to put ourselves in his place, and to live through his experiences when he identifies and believes without question that his words 'are' the things they only stand for. To give the full consequences of such identification resulting in wrong evaluation, I might add most tedious *descriptions* of the interplay of situations, evaluations., in quarrels, unhappinesses, disagreements., leading to dramas and tragedies, as well as to many forms of 'mental' illness effectively described only in the *belles-lettres*. Thus, Smith[1], who is *not* conscious of abstracting, makes the statement, 'A circle is not square'. Let us suppose that Brown[1] contradicts him. Smith[1] is angered; for his *s.r*, his statement 'is' the 'plain truth', and Brown[1] must be a fool. He objectifies it, ascribes to it undue value. For him, it 'is' 'experience', a 'fact'., and he bursts into speech, denouncing Brown[1] and showing how wrong he 'is'. From this semantic attitude, many difficulties and tragedies arise.

But if Smith[2] (conscious of abstracting) makes the statement, 'A circle is not square', and Brown[2] contradicts him, what would Smith[2] do? He would smile, would not burst into speech to defend *his* statement, but would ask Brown[2], 'What do you mean? I do not quite understand you'. After receiving some answer, Smith[2] would explain to Brown[2] that his statement is not anything to quarrel about, as it is verbal and is true only by *definition*. He would also grant the right of Brown[2] *not* to accept *his definition*, but to use another one to satisfy himself. The problem would then, naturally, arise as to what definition both could accept, or which would be generally acceptable. And the problem would then be solved by purely pragmatic considerations. Words appear as creatures of definitions, and optional; but this attitude involves important and new *s.r.*

This fact seems of tremendous semantic importance, as it provides the working foundation for a theory of 'universal agreement'. In the first part of the above example, Smith[1], according to the accepted standards, was 'right' ('a circle is not square'). Is he 'more right' than Brown[1], for whom the 'circle is square'? Not at all. Both statements belong to the verbal level, and represent only forms of representation for *s.r inside their skin*. Either may be 'right' by some explicit or implicit 'definitions'. Are the two statements equally valid? This *we do not know a priori;* we must investigate to find out if the noises uttered have meanings outside of pathology, or which statement structurally covers the situation better, carries us structurally further in describing and analysing this world, . Only scientific structural analysis can give the preference to one form over another. Smith and Brown can only produce their 'definitions' according to their *s.r*, but they are *not* judges as to which 'definitions' will *ultimately* stand the test of structure.

The moment we eliminate identification we become conscious of abstracting, and permanently and instinctively remember that the object is *not* the event, that the label is *not* the object, and that a statement about a statement is *not* the first statement; thus, we reach a semantic state, where we recognize that everybody 'is right' by his own 'definitions'. But any individual or unenlightened public opinion is not the sole judge as to what 'definitions' and what language should prevail. Only structural investigation (science) can decide which appears as the structurally more similar form of representation on the verbal levels for what is going on on the un-speakable, objective levels.

When it comes to 'description of facts', the situation is not fundamentally altered. Mistakes seem always possible and often occur. Besides, the semantic impressions which 'facts' make on us are also individual, and often in conflict, as comparison of the testimonies of eye-witnesses shows. But there is no need for permanent disagreement; more structural investigation of the objective and verbal levels will provide a solution. Once such an investigation is carried far enough, we can always reach a semantic basis where all may agree, provided we do not identify, do not objectify, and do not confuse description and inference, descriptive and inferential words, .

As our analysis is carried out from the structural and *non-el* point of view, we should not miss the fact that semantic components associated with words and statements are, outside of very pathological cases, never entirely absent, and become of paramount importance. In the older days, we had no simple and effective means by which we could affect painful, misplaced, or disproportionate evaluations, meanings., through a semantic re-education, which are supplied by the present analysis and the use of the Structural Differential. The means to eliminate identification consist of: first, an *objective* relief diagram to which we can *point our finger;* and second, a convincing explanation (pointing the finger to the labels) that the verbal levels, with their distressing and disastrous older *s.r, are not,* and differ entirely from, the levels of objects and events. Whatever we may say or feel, the objects and events remain on the un-speakable levels and cannot be reached by words. Under such natural structural conditions, we can only reach the objective level by seeing, handling, actually feeling., and, therefore, by pointing our finger to the object on the Structural Differential and being silent—all of which cannot be conveyed by words *alone.*

In experiments with the 'mentally' ill in whom the semantic disturbances were very strong, it took several months to train the patients in non-identity and in silence on the objective levels. But, as soon as this was achieved, either complete or partial relief followed.

The main disturbances in daily life, as well as in 'mental' illness, are found in the affective field. We find an internal pressure of identifications, expressed by bursting into speech, and unjustified semantic over-evaluation of words, the ascribing of objectivity to words, . In such cases, suppression or repression of words does not accomplish much, but often does *considerable harm* and must be avoided by all means. Under such conditions, the use of the relief diagram becomes a necessity in pointing to the difference between different orders of abstractions and inducing the semantically beneficial silence on the 'objective level' without repression or suppression.

With the use of the Structural Differential, we can eliminate identification, and so attain the benefits, avoiding the dangers. If any one identifies, and his *s.r* drive him into an outburst of speech, we do not repress or suppress him; we say, instead: 'At your pleasure [since it makes him feel better], but remember that your words occur on the verbal levels [showing with a gesture of the hands the hanging labels], and that they *are not* the objective level, which remains untouched and unchanged'. Such a procedure, when repeated again and again, gives him the proper semantic *evaluation of orders of abstractions, frees him from identification, yet without repression or suppression.* It teaches him, also, to enquire into alleged 'facts', and then to try to find structurally better forms of representation. If such results are not forthcoming, we may use the older forms, but by proper evaluation we do not semantically put 'belief' in these forms of representation. Such beliefs always appear as the result of identification somewhere.

The technique of training is simple. We live on the 'objective' or lower order of abstraction levels, where we must see, feel, touch, *perform.,* but *never* speak. In training, we must use our hands, . It is very useful, after the Structural Differential has been repeatedly explained, stressing, in particular, the rejection of the 'is' of identity, not to interrupt the other fellow. Let him speak, but wave the hand, indicating the verbal levels; then point the finger to the objective level, and with the other hand, close your own lips, to show that on the objective level one can only be silent. When performed repeatedly, this pantomime has a most beneficial, semantic, pacifying effect upon the 'over-emotionalized' identification-conditions. The neurological mechanism of this action is not fully known, but some aspects are quite clear.

The more elaborate a nervous system becomes, the further some parts of the brain are removed from immediate experience. Nerve currents, having finite velocity, eventually have longer and more numerous paths to travel; different possibilities and complications arise, resulting in 'delayed action'. It is known that the thalamus (roughly) appears connected with affective and 'emotional' life, and that the cortex, farther removed and isolated from the external world, has the effect of inducing this 'delay in action'. In unbalanced and 'emotional' 'thinking', which is so prevalent, the thalamus seems overworked, the cortex seems not worked enough. The results take on the form of a low kind of animalistic, primitive, or infantile behaviour, often of a pathological character in a supposedly civilized adult. It appears that the 'silence on objective levels' introduces this 'delayed action', unloading the thalamic material on the cortex. This psychophysiological method is very simple, scientific, and entirely general. The standard 'mental' therapy of today applies also a *method of re-education* of s.r, as if relieving the thalamus, and putting more of the nerve currents through the cortex, or eventually furnishing the cortex with different material, so that the thalamic material returning from the cortex could be properly influenced.

If we succeed in such a semantic re-education, the difficulties vanish. The older experimental data show that in many instances we have succeeded, and that in many we have failed. The successful cases show that we actually know the essential semantic points involved; the failures show that we do not know enough, and that our older theories are not sufficiently general. At present, only the more pronounced and morbid semantic disturbances come to the attention of physicians, and very little is done by way of *preventive* measures. Besides the pronounced disturbances in daily life, we see an enormous number of semantic disturbances which we disregard, and call 'peculiarities'. In the majority of cases, these 'peculiarities' are undesirable, and, under unfavorable conditions, may lead to more serious consequences of a morbid character. They usually involve a great deal of unhappiness for all concerned, and unhappiness appears as a sign of some semantic maladjustment somewhere, and so may be destructive to 'mental' and nervous health.

In advanced 'mental' illnesses, such as usually come to the attention of psychiatrists, there are certain psycho-logical symptoms which are generally present. The symptoms of interest to us in this work are called 'delusions', 'illusions', and 'hallucinations'. All of them involve the semantic identification or *confusions of the orders of abstractions*, the evaluation of lower orders of abstraction as higher, or higher as lower. It was explained already that some components of identification are invariably present there, and so identification may be considered as an elementary type of semantic disturbances from which all the other states differ only in intensity.

The main point is to find psychophysiological preventive means whereby this identification can be forestalled or eliminated. To date, experience and analysis show that all forms of identification may be successfully eliminated by training in *visualization*, if this semantic state can be produced. For this purpose the Structural Differential is uniquely useful and necessary. With its help we train all centres. The lower centres are involved, as we see, feel, hear.; the higher centres are equally involved, as we 'remember', 'understand'.; *with the result that all centres work together without conflict*. The 'consciousness of abstracting' is inculcated, replacing vicious s.r of confusion of orders of abstractions and identification.

This harmonious working of all centres on their proper levels has extremely far-reaching, practical consequences in 'mental' and physical hygiene. We become co-ordinated, adjusted, and difficulties which might otherwise occur in the future are eliminated in a preventive way. It must be remembered that, at present, it is impossible to foresee to how great an extent the elimination of identification on all levels will have a beneficial effect. At this stage we know even experimentally that the benefits are very large, but we may expect that they will become still more numerous when more experimenting has been done. Delusions, illusions, and hallucinations represent manifestations which occur in practically all 'mental' difficulties, and they only represent a semantic identification of orders of abstractions of different degrees of intensity. When this confusion is eliminated, we may expect general changes in the symptoms. But as the correspondence is probably not *one-to-one*, it is impossible to foretell theoretically what improvements may be expected in pronounced illness. In the slighter disturbances, which affect us in daily life, the results are much easier to foresee, and are *always* beneficial.

To how great an extent the consciousness of abstracting benefits semantically *the whole organism*, I may illustrate by one of my own experiences. Once I was travelling on a ship. A gentleman visited my cabin, and, seeing the Structural Differential, asked questions about it. After a short explanation, he asked about practical applications.

My guest was sitting on my berth; I was sitting on a small folding chair. I got up, went to the door, then pretended that I was coming in, and, at my suggestion, he said, 'Please have a seat'. I remained standing while explaining how, if I were not 'conscious of abstracting', to me his word 'seat' would be identified with the chair (objectification) and my *s.r* would be such that I would sit down with great confidence. If the chair were to collapse I would have, besides the bump, an affective shock, 'fright'., which might do harm to my nervous system. But if I were conscious of abstracting, my *s.r* would be different. I would remember that the *word*, the *label* 'seat' is *not* the thing on which I am supposed to sit. I would remember that I am to sit on this individual, unique, un-speakable object, which might be strong or weak, . Accordingly, I would sit carefully. In case the chair should collapse, and I should hurt myself physically, I would still have been saved an affective nervous shock.

During all these explanations I was handling the little chair and shaking it. I did not notice that the legs were falling out, and that the chair was becoming unfit for use. Then, when I actually sat on the relic, it gave way under me. However, I did *not fall* on the floor. I caught myself in the air, so to say, and saved myself from a painful experience. It is important to notice that such physical readiness requires a very elaborate, nervous, unconscious co-ordination, which was accomplished by the semantic state of *non-identification* or *consciousness of abstracting*. When such a consciousness of abstracting is acquired, it works instinctively and automatically and does *not* require continual effort. Its operation involves a fraction of a second's *delay in action*, but this small delay is not harmful in practice; on the contrary, it has very important psycho-logical and neurological 'delayed action' effects.

It seems that 'silence' on the objective levels involves this psychophysiological delay. No matter how small, it serves to unload the thalamic material on the cortex. In a number of clinical cases, Dr. Philip S. Graven has demonstrated that the moment such a delay can actually be produced in the patient, he either improves or is entirely relieved. The precise neurological mechanism of this process is not known, but there is no doubt that this 'delayed action' has many very beneficial effects upon the whole working of the nervous system. It somehow balances harmful *s.r*, and also somehow stimulates the higher nervous centres to more *physiological* control over the lower centres.

A very vital point in this connection should be noticed. That this 'delayed action' is beneficial is acknowledged by the majority of normally developed adults in the form of delay in action and finds its expression in such statements as 'think twice', 'keep your head', 'hold your horses', 'keep cool', 'steady', 'wait a minute'., and such functional recipes as 'when angry, count ten', . In daily life, such wisdom is acquired either by painful experience, or is taught to children in an [A] language, which, as practice shows, is rarely effective because of its inadequacy. It is seldom realized that the mechanism of these functional observations and familiar advices have very powerful and workable underlying *neurological processes*, which can be *reached and directly affected by psychophysiological*, ordinal, *non-el* methods in connection with *the structure of the language we use*. Thus, under an infantile, [A], and prevailing system we use and teach our children a language involving the 'is' of identity, and so we must confuse orders of abstractions, preparing for ourselves and the children the harmful semantic predispositions for 'bursting into speech', instead of 'wait a minute', which, neurologically, means abusing our thalamus and keeping our cortex 'unemployed'. In a [non-A] ∞-valued system we reject the 'is' of identity; we cannot confuse orders of abstractions; we cannot identify words with the un-speakable objective levels or inferences with descriptions, and we cannot identify the different abstractions of different individuals, . This semantic state of proper evaluation results in discrimination between the different orders of abstractions; an automatic delay is introduced—the cortex is switched completely into the nervous circuit. The semantic foundation is laid for 'higher mentality' and 'emotional balance'.

We have already had occasion to mention the mechanism of projection in connection with identification, as a semantic state of affective ascribing of lower centre characteristics to higher order abstractions and vice versa, and in connection with the introverted or extroverted attitudes. Likewise, we have already reached the conclusion that a well-adjusted and, therefore, well-balanced individual should be neither of the extremes, but a balanced extroverted introvert. By training with the Differential this important semantic result may be brought about. By training with the 'object' on its level, we become extroverted, and we learn to observe; this results in semantic freedom from 'preconceived ideas', such as we have when we start with the evaluation, label first and object next, instead of the natural order, object first and label next. By passing to higher order abstractions and evaluating the successive ranks of labels, we train in introversion. The result, as a whole, is that we may achieve the desirable and balanced semantic state of the extroverted introvert.

That in the training with the Differential we use all available nerve centres is beneficial, because the lower centres are in closer connection with the vegetative nervous system than are the others.

15. HIGHER ORDER ABSTRACTIONS

General

In the previous chapters I demonstrated that there is a short cut which enables us to grasp, acquire, and apply what has been advanced in the present work. This semantic short cut is 'consciousness of abstracting'. It is a psycho-logical attitude toward all our abstracting on all levels, and so involves the co-ordinated working of the organism-as-a-whole.

The use of the Structural Differential is necessary, because some levels are un-speakable. We can see them, handle them, feel them., but under *no* circumstances can we reach those levels by speech alone. We must, therefore, have a diagram, by preference in relief form, which represents the empirical structural conditions, and which indicates the un-speakable level by some other means than speech. We must, in the simplest case, either point our finger to the object, insisting upon silence, or must perform some bodily activity and similarly insist upon silence, as the performing and feelings are also *not* words.

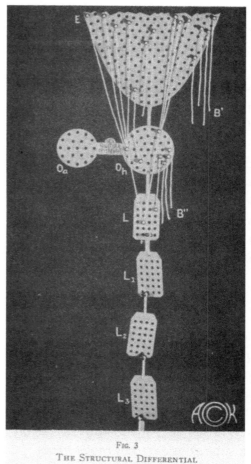

FIG. 3
THE STRUCTURAL DIFFERENTIAL

In such semantic training it is enough to insist upon the non-identity or the difference between the objective, *un-speakable* levels of lower order abstractions, (O_h), and the verbal or higher order abstractions, (L_n). When this habit and feeling are acquired, no one should have difficulties in extending the non-identity method to daily-life occurrences. To achieve these semantic aims, we must first emphasize the common-sense fact that an object *is not* the event. To do this, we start with the 1933 scientific structural 'metaphysics' about the event and stress the fact *that* the object, being a nervous abstraction of lower order, has fewer and different m.o characteristics than the event has. This is best accomplished by stressing the fact that in abstracting from the event to the object we left out some characteristics. We did not *abstract 'all'* characteristics; this would be a self-contradiction in terms, an impossibility.

We do not even need to stress a full understanding of the event. Common-sense examples, showing that what we recognize as a 'pencil' is not 'all', often suffice. No one will have difficulties, provided he trains himself in this direction, in *remembering* continually and instinctively the free hanging strings (B'), (B''), which indicate the non-abstracted or left-out characteristics and which help to train in *non-identity*. With the relief diagram, the s.r of the student are trained *through all nervous centres*. He sees, he handles., the hanging strings, and he also hears about them. This gives the *maximum probability* that the organism-as-a-whole will be affected. In this way an 'intellectual' theory engages the 'senses', feelings, and reflex mechanisms. To affect the organism-as-a-whole, organism-as-a-whole methods must be employed.

A similar structural situation is found when we deal with higher order abstractions. A word, or a name, or a statement is conveyed in spoken form or by writing, and affects first the lower centres and then is abstracted, and again transformed, by the higher centres. The order is generally not changed when the verbal issues are neither seen nor heard but originate in ourselves. Most 'impulses', 'interests', 'meanings', 'evaluations'., originate in lower centres and follow the usual course, from lower centres to higher. When 'experience' (reaction of lower centres) is transformed into 'memories' (higher centres)., the order is similar. Difficulties begin when the order is pathologically *reversed* and 'ideas' are evaluated as experience, words as objects, . In the building of language a similar process can be observed. We observe the absolute individuals with which we actually deal, we label them with individual names, say, A_1, A_2, . . ., A_{11}, A_{12}, . . ., A_{21}, A_{22}, . . ., A_{31}, A_{32}, . . ., . By a process of abstracting and *disregarding*, for instance, the characteristic subscripts '1', we would have only the ones which have the characteristic subscripts 2, 3, . . ., 9, 22, 23, . . ., 29, . *Disregarding* the characteristic subscripts '2', we would have the ones with the characteristics 3, 4, ... 9, 33, 34 . . . ., 39, . Finally, if we should eliminate all *individual* characteristic subscripts, we would have a 'general' name A for the whole group without singling out individual characteristics.

All words of the type of 'man', 'animal', 'house', 'chair', 'pencil'., have been built by a similar process of abstraction, or *disregard* for individual differences. In each case of disregard of individual characteristics a *new* neurological process was involved.

Similarly, with 'statements about a statement'. When we hear a statement, or see it in a written form, such a statement becomes a stimulus entering through the lower centres, and a statement about it represents, in general, a new process of abstraction, or an abstraction of higher order.

It becomes obvious that the introduction of a language of 'different order of abstractions', although it is not familiar, yet structurally it represents very closely, in terms of *order*, most fundamental neurological processes going on in us. As we already know, a natural order has been established by evolution; namely, lower order abstractions first, higher next; the identifications of orders or the reversal of orders appears pathological for man and appears as a confusion of orders of abstractions, resulting in *false evaluation:* identification, illusions, delusions, and hallucinations.

Multiordinal Terms

In the examples given in Section A, we used words such as 'proposition', which were applied to all higher order abstractions. We have already seen that such terms may have different uses or meanings if applied to different orders of abstractions. Thus originates what I call the *multiordinality* of terms. The words 'yes', 'no', 'true', 'false', 'function', 'property', 'relation', 'number', 'difference', 'name', 'definition', 'abstraction', 'proposition', 'fact', 'reality', 'structure', 'characteristic', 'problem', 'to know', 'to think', 'to speak', 'to hate', 'to love', 'to doubt', 'cause', 'effect', 'meaning', 'evaluation', and an endless array of the most important terms we have, must be considered as *multiordinal terms*. There is a most important semantic characteristic of these *m.o* terms: namely, that they are ambiguous, or ∞-valued, in general, and that each has a definite meaning, or one value, only and exclusively in a given context, when the order of abstraction can be definitely indicated.

These issues appear extremely simple and general, a part and parcel of the structure of 'human knowledge' and of our Language. We cannot avoid these semantic issues, and, therefore, the only way left is to face them explicitly. The test for the multiordinality of a term is simple.

Let us make any statement and see if a given term applies to it ('true', 'false', 'yes', 'no', 'fact', 'reality', 'to think', 'to love', .). If it does, let us deliberately make another statement *about* the former statement and test if the given term may be used again. If so, it is a safe assertion that this term should be considered as *m.o*. Any one can test such a *m.o* term by himself without any difficulty. The main point about all such *m.o* terms is that, *in general*, they are *ambiguous*, and that all arguments about them, 'in general', lead only to *identification of orders of abstractions and semantic disturbances, and nowhere else.* Multiordinal terms have only definite meanings on a given level and in a given context. Before we can argue about them, we must fix their orders, whereupon the issues become simple and lead to agreement. As to 'orders of abstraction', we have no possibility of ascertaining the 'absolute' order of an abstraction; besides, we *never* need it. In human semantic difficulties, in science, as well as in private life, usually no more than three, perhaps even two, neighbouring levels require consideration. When it comes to a serious discussion of some problem, errors, ambiguity, confusion, and disagreement follow from confusing or identifying the neighbouring levels. In practice, it becomes *extremely simple* to settle these three (or two) levels and to keep them separated, *provided we are conscious of abstracting, but not otherwise.*

For a theory of sanity, these issues seem important and structurally essential. In identifications, delusions, illusions, and hallucinations, we have found a *confusion* between the orders of abstractions or a false evaluation expressed as a reversal of the natural order.

One of the symptoms of this confusion manifests itself as 'false beliefs', which again imply comparison of statements about 'facts' and 'reality', and involve such terms as 'yes', 'no', 'true', 'false', . As all these terms are multiordinal, and, therefore, ambiguous, 'general' 'philosophical' rigmaroles should be avoided. With the consciousness of abstracting, and, therefore, with a *feel* for this peculiar stratification of 'human knowledge', all semantic problems involved can be settled simply.

The avoidance of *m.o* terms is impossible and undesirable. Systematic ambiguity of the most important terms follows systematic analogy. They appear as a direct result and condition of our powers of abstracting in different orders, and allow us to apply one chain of ∞-valued reasoning to an endless array of different one-valued facts, all of which are different and become manageable only through our abstracting powers.

The semantic benefits of such a recognition of multiordinality are, in the main, seven-fold: (1) we gain an enormous economy of 'time' and effort, as we stop 'the hunting of the snark', usually called 'philosophy', or for a one-valued general definition of a *m.o* term, which would not be formulated in other *m.o* terms; (2) we acquire great versatility in expression, as our most important vocabulary consists of *m.o* terms, which can be extended indefinitely by assigning many different orders and, therefore, meanings; (3) we recognize that a definition of a *m.o* term must, by necessity, represent not a proposition but a propositional function involving variables; (4) we do not need to bother much about formal definitions of a *m.o* term outside of mathematics, but may use the term freely, realizing that its unique, in principle, meaning in a given context is structurally indicated by the context; (5) under such structural conditions, the freedom of the writer or speaker becomes very much accentuated; his vocabulary consists potentially of infinite numbers of words, and psycho-logical, semantic blockages are eliminated; (6) he knows that a reader who understands that ∞-valued mechanism will never be confused as to the meaning intended; and (7) the whole linguistic process becomes extremely flexible, yet it preserves its essential extensional one-valued character, in a given case.

In a certain sense, such a use of *m.o* terms is to be found in poetry, and it is well known that many scientists, particularly the creative ones, like poetry. Moreover, poetry often conveys in a few sentences more of lasting values than a whole volume of scientific analysis. The free use of *m.o* terms without the bother of a structurally impossible formalism outside of mathematics accomplishes this, *provided we are conscious of abstracting; otherwise only confusion results.*

It should be understood that I have no intention of condemning formalism. Formalism of the most rigorous character is an extremely important and valuable discipline (mathematics at present); but formalism, as such, in experimental science and life appears often as a handicap and not as a benefit, because, in empirical science and life, we are engaged in exploring and discovering the unknown structure of the world as a means for structural adjustment. The formal elaboration of some language is only the consistent elaboration of its structure, which must be accomplished independently if we are to have means to compare verbal with empirical structures. From a [non-A] point of view, both issues are equally important in the search for structure.

Under such structural empirical conditions the *m.o* terms acquire great semantic importance, and perhaps, without them, language, mathematics, and science would be impossible. As soon as we understand this, we are forced to realize the profound structural and semantic difference between the [A] and [non-A] systems. What in the old days were considered propositions, become propositional functions, and most of our doctrines become the doctrinal functions of Keyser, or system-functions, allowing multiple interpretations.

Terms belong to verbal levels and their meanings *must* be given by definitions, these definitions depending on undefined terms, which consist always, as far as my knowledge goes, of *m.o* terms. Perhaps it is necessary for them to have this character, to be useful at all. When these structural empirical conditions are taken into account, we must conclude that the postulational method which gives the structure of a given doctrine lies at the foundation of all human linguistic performances, in daily life as well as in mathematics and science. The study of these problems throws a most important light on all mysteries of language, and on the proper use of this most important human neurological and semantic function, without which sanity is impossible.

From a structural point of view, postulates or definitions or assumptions must be considered as those relational or multi-dimensional order structural assumptions which establish, conjointly with the undefined terms, the structure of a given language. Obviously, to find the structure of a language we must work out the given language to a system of postulates and find the minimum of its (never unique) undefined terms. This done, we should have the structure of such a system fully disclosed; and, with the structure of the language thoroughly known, we should have a most valuable tool for investigating empirical structure by predicting verbally, and then verifying empirically.

To pacify the non-specialist, let me say at once that this work is very tedious and difficult, although a crying need; nevertheless, it may be accomplished by a single individual. Because of the character of the problem, however, when this work is done, the semantic results have always proved thus far—and probably will continue so—quite simple and comprehensible to the common sense, even of a child.

One very important point should be noted. Since language was first used by the human race, the structural and related semantic conditions disclosed by the present analysis *have not been changed*, as they are inherent in the structure of 'human knowledge' and language. Historically, we were always most interested in the immediacy of our daily lives. We began with grunts symbolizing this immediacy, and we never realize, even now, that these historically first grunts were the most complex and difficult of them all. Besides these grunts, we have also developed others, which we call mathematics, dealing with, and elaborating, a language of numbers, or (as I define it semantically) a language of *two symmetrical and infinitely many asymmetrical unique, specific relations* for exploring the structure of the world, which is, at present, the most effective and the simplest language yet formed. Only in 1933, after many hundreds of thousands of years, have the last mentioned grunts become sufficiently elaborate to give us a sidelight on structure. We must revise the whole linguistic procedure and structure, and gain the means by which to disclose the structure of 'human knowledge'. Such semantic means will provide for the proper handling of our neurological structure, which, in turn, is the foundation for the structurally proper use of the human nervous system, and will lead to human nervous adjustment, appropriate *s.r*, and, therefore, to sanity.

Human beings are quite accustomed to the fact that words have different meanings, and by making use of this fact have produced some rather detrimental speculations, but, to the best of my knowledge, the structural discovery of the multiordinality of terms and of the psycho-physiological importance of the treatment of orders of abstractions resulting from the rejection of the 'is' of identity—as formulated in the present system—is novel. In this mechanism of multiordinality, we shall find an unusually important structural problem of human psychologics, responsible for a great many fundamental, desirable, undesirable, and even morbid, human characteristics. The full mastery of this mechanism is only possible when it is formulated, and leads automatically to a possibility of a complete psychophysiological adjustment. This adjustment often reverses the psycho-logical process prevailing at a given date; and this is the foundation, among others, of what we call 'culture' and 'sublimation' in psychiatry.

Let me recall that one of the most fundamental functional differences between animal and man consists in the fact that no matter in how many orders the animal may abstract, its abstractions stop on some level beyond which the animal cannot proceed. Not so with man. Structurally and potentially, man can abstract in indefinitely many orders, and no one can say legitimately that he has reached the 'final' order of abstractions beyond which no one can go. In the older days, when this semantic mechanism was not made structurally obvious, the majority of us copied animals, and stopped abstracting on some level, as if this were the 'final' level. In our semantic training in language and the 'is' of identity given to us by our parents or teachers or in school, the multiordinality of terms was never suspected, and, although the human physiological mechanism was operating all the while, we used it on the conscious level in the animalistic way, which means ceasing to abstract at some level. Instead of being told of the mechanism, and of being trained consciously in the fluid and dynamic *s.r* of *passing to higher and higher abstractions as normal*, for Smith, we preserved a sub-normal, animalistic semantic blockage, and 'emotionally' stopped abstracting on some level.

Thus, for instance, if as a result of life, we come to a psychological state of hate or doubt, and stop at that level, then, as we know from experience, the lives of the given individual and of those close to him are not so happy. But a hate or doubt of a higher order reverses or annuls the first order semantic effect. Thus, hate of hate, or doubt of doubt—a second order effect—has reversed or annulled the first order effect, which was detrimental to all concerned because it remained a *structurally-stopped or an animalistic* first order effect.

The whole subject of our human capacity for higher abstracting without discernible limits appears extremely broad, novel, and unanalysed. It will take many years and volumes to work it out; so, of necessity, the examples given below will be only suggestive and will serve to illustrate roughly the enormous power of the [non-A] methods and structure, aiming to make them workable as an educational, powerful, semantic device.

Let us take some terms which may be considered as of a positive character and represent the structure of 'culture', science, and what is known in psychiatry as 'sublimation'; such as curiosity, attention, analysis, reasoning, choice, consideration, knowing, evaluation, . The first order effects are well known, and we do not need to analyse them. But if we transform them into second order effects, we then have curiosity of curiosity, attention of attention, analysis of analysis, reasoning about reasoning' (which represents science, psycho-logics, epistemology.,); choice of choice (which represents freedom, lack of psycho-logical blockages, and shows, also, the semantic mechanism of eliminating those blocks); consideration of consideration gives an important cultural achievement; knowing of knowing involves abstracting and structure, becomes 'consciousness', at least in its limited aspect, taken as consciousness of abstracting; evaluation of evaluation becomes a theory of sanity, .

Another group represents morbid semantic reactions. Thus the first order worry, nervousness, fear, pity., may be quite legitimate and comparatively harmless. But when these are of a higher order and identified with the first order as in worry about worry, fear of fear., they become morbid. Pity of pity is dangerously near to self-pity. Second order effects, such as belief in belief, makes fanaticism. To know that we know, to have conviction of conviction, ignorance of ignorance., shows the mechanism of dogmatism; while such effects as free will of free will, or cause of cause., often become delusions and illusions.

A third group is represented by such first order effects as inhibition, hate, doubt, contempt, disgust, anger, and similar semantic states; the *second order reverses and annuls* the first order effects. Thus an inhibition of an inhibition becomes a positive excitation or release (see Part VI); hate of hate is close to 'love'; doubt of doubt becomes scientific criticism and imparts the scientific tendency; the others obviously reverse or annul the first order undesirable *s.r.*

In this connection the pernicious effect of identification becomes quite obvious. In the first and third cases beneficial effects were *prevented*, because identification of orders of abstractions, as a semantic state, produced a semantic blockage which did not allow us to pass to higher order abstractions; in the second ease, it actually produced morbid manifestations.

The consciousness of abstracting, which involves, among others, the full instinctive semantic realization of non-identity and the stratification of human knowledge, and so the multiordinality of the most important terms we use, solves these weighty and complex problems because it gives us structural methods for semantic evaluation, for orientation, and for handling them. By passing to higher orders these states which involve inhibition or negative excitation become reversed. Some of them on higher levels become culturally important; and some of them become morbid. Now consciousness of abstracting in all cases gives us the semantic *freedom* of all levels and so helps *evaluation* and selection, thus removing the possibility of remaining animalistically fixed or blocked on any one level. Here we find the mechanism of the 'change of human nature' and an assistance for persons in morbid states to revise by themselves their own afflictions by the simple realization that the symptoms are due to identifying levels which are essentially different, an unconscious jumping of a level or of otherwise confusing the orders of abstractions. Even at present all psychotherapy is unconsciously using this mechanism, although, as far as I know, it has never before been structurally formulated in a general way.

It should be added that the moment we eliminate identification and acquire the consciousness of abstracting, as explained in the present system, we have already acquired the permanent semantic feeling of this peculiar *structural stratification* of human knowledge which is found in the psycho-logics of the differential and integral calculus and mathematics, similar in structure to the world around us, without any difficult mathematical technique. Psycho-logically, both mathematics and the present system appear structurally similar, not only to themselves, but also to the world and our nervous system; and at this point it departs very widely from the older systems.

Confusion of Higher Orders of Abstraction

We have already seen that Fido's power of abstracting stops somewhere. If we are finalists of any kind, we also assume that *our* power of abstracting stops somewhere. In some such way the finalistic, dogmatic and absolutistic semantic attitudes are built.

If, however, by the aid of the Structural Differential we train the *s.r* of our children in [non-A] non-identity and the inherent stratification of human knowledge and power of abstracting, we *facilitate* the passing to higher order abstractions and establish *flexible s.r* of *full conditionality* which are unique for Smith and of great preventive and therapeutic value. We thus build up 'human mind' for efficiency and sanity, by eliminating the factors of semantic blockages, while, by engaging the activity of the higher nerve centres, we diminish the vicious overflow of nervous energy upon the lower nerve centres, which, if allowed, must, of necessity, make itself manifest in arrested or regressive symptoms.

The above issues are of serious semantic importance in our daily lives and in sanity. All semantic disturbances involve evaluation, doctrines, creeds, speculations., and vice versa. Under circumstances such as described above, which appear inherent with us, it is dangerous not to have means to see one's way clear in the maze of verbal difficulties with all their dangerous and ever-present semantic components.

By disregarding the orders of abstractions, we can manufacture any kind of verbal difficulties; and, without the consciousness of abstracting, we all become nearly helpless and hopeless semantic victims of a primitive-made language and its underlying structural metaphysics. Yet the way out is simple; non-identity leads to 'consciousness of abstracting' and gives us a new working sense for *values*, new *s.r*, to guide us in the verbal labyrinth.

Outside of 'objectification', which is defined as the evaluation of higher order abstractions as lower; namely, words, memories., as objects, experiences, feelings., the most usual identification of different *higher order* abstractions appears as the confusion of inferences and inferential terms with descriptions and descriptive terms.

Obviously, if we consider a description as of the *n*th order, then an inference from such a description (or others) should be considered as an abstraction of a higher order ($n+1$). Before we make a decision, we usually make a more or less hasty survey of happenings, this survey establishing a foundation for our judgements, which become the basis of our action. This statement is fairly general, as the components of it can be found by analysis practically everywhere. Our problem is to analyse the general case. Let us follow up roughly the process.

We assume, for instance, an hypothetical case of an ideal observer who observes correctly and gives an impersonal, unbiased account of what he has observed. Let us assume that the happenings he has observed appeared as: ▓, ▶, ▫, ☼, ..., and then a new happening ⫴ occurred. At this level of *observation*, no speaking can be done, and, therefore, I use various fanciful symbols, and not words. The observer then gives a *description* of the above happenings, let us say *a, b, c, d, . . ., x*; then he makes an inference from these descriptions and reaches a conclusion or forms a judgement *A* about these facts. We assume that facts unknown to him, which always exist, are not important in this case. Let us assume, also, that his conclusion seems correct and that the action *A''* which this conclusion motivates is appropriate. Obviously, we deal with at least three different levels of abstractions: the seen, experienced., lower order abstractions (unspeakable); then the descriptive level, and. finally, the inferential levels.

Let us assume now another individual, Smith₁, ignorant of structure or the orders of abstractions, of consciousness of abstracting, of *s.r.*; a politician or a preacher, let us say, a person who habitually identifies, confuses his orders, uses inferential language for descriptions, and rather makes a business out of it. Let us assume that Smith₁ observes the 'same happenings'. He would witness the happenings ▓, ▶, ◻, ☼, ..., and the happening 卌 would appear new to him. The happenings, ▓, ▶, ◻, ☼, ..., he would describe in the form *a, b, c, d, . . .*, from which fewer descriptions he *would form a judgement, reach a conclusion, B;* which means that he would pass to another order of abstractions. When the new happening 卌 occurs, he handles it with an already formed opinion *B*, and so his description of the happening 卌 is *coloured* by his older *s.r* and no longer the *x* of the ideal observer, but *B(x)=y*. His description of 'facts' would *not* appear as the *a, b, c, d, . . ., x,* of the ideal observer but *a, b, c, d, . . ., B(x)=y.* Next he would abstract on a higher level, form a new judgement, about 'facts' *a, b, c, d, . . ., B(x)=y,* let us say, *C.* We see how the semantic error was produced. The happenings appeared the 'same', yet the unconscious identification of levels brought finally an entirely different conclusion to motivate a quite different action, *C''*.

A diagram will make this structurally clearer, as it is very difficult to explain this by words alone. On the Structural Differential it is shown without difficulty.

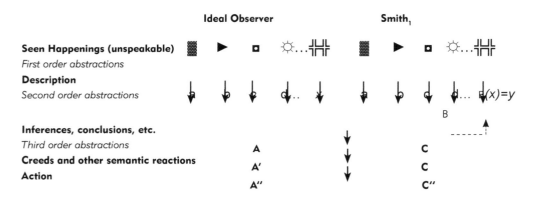

	Ideal Observer	Smith₁
Seen Happenings (unspeakable) *First order abstractions*	▓ ▶ ◻ ☼...卌	▓ ▶ ◻ ☼...卌
Description *Second order abstractions*	↓ ↓ ↓ ↓.. ↓	↓ ↓ ↓ ↓... *B(x)=y*
Inferences, conclusions, etc. *Third order abstractions*	A	C
Creeds and other semantic reactions	A'	C
Action	A''	C''

Let us illustrate the foregoing with two clinical examples. In one case, a young boy persistently did not get up in the morning. In another case, a boy persistently took money from his mother's pocketbook. In both cases, the actions were undesirable. In both cases, the parents unconsciously identified the levels, *x* was identified with *B(x)*, and confused their orders of abstractions. In the first case, they *concluded* that the boy was *lazy*; in the second, that the boy was a *thief.* The parents, through semantic identification, read these inferences into every new 'description' of forthcoming facts, so that the parents' new 'facts' became more and more semantically distorted and coloured in evaluation, and their actions more and more detrimental to all concerned. The general conditions in both families became continually worse, until the reading of inferences into descriptions by the ignorant parents produced a semantic background in the boys of driving them to murderous intents.

A psychiatrist dealt with the problem as shown in the diagram of the ideal observer. The net result was that the one boy was not 'lazy', nor the other a 'thief', but that both were ill. After medical attention, of which the first step was to clarify the symbolic semantic situation, though not in such a general way as given here, all went smoothly. Two families were saved from crime and wreck.

Let us follow our *daily experiences* by the aid of the Structural Differential. We find ourselves on at least five levels. The first represents the un-speakable event, or the scientific object, or the unseen physico-chemical processes on the sub-microscopic levels which constitute stimuli registered by our nervous system as objects. The second consists of the external, objective, also un-speakable, levels on which we see with our eyes, . On this level, we could make a moving picture, including actions., (writing a book is also behaviour). The third level represents the equally un-speakable psycho-logical 'pictures' and *s.r.* On the fourth level of abstractions we describe verbally our facts, that humans (*a*) eat, sleep.; (*b*) cheat, murder.; (*c*) moralize, philosophize, legislate.; (*d*) scientize, mathematize, . Finally, in the present context, our inferences belong to the fifth level.

Unfortunately, we usually abstract facts (*a*), identify the levels, and form a conclusion 'man is an animal', . From this *conclusion* we confuse the levels again and colour the description of the facts (*b*), (*c*), (*d*).; jump again to higher levels and build conclusions from descriptions (*a*) and from *distorted*, coloured descriptions (*b*), (*c*), (*d*), and so obtain the prevailing doctrines in all fields. These again lead us, in the field of action, to the mess we all find ourselves in. In this dervish dance between the levels we entirely *disregarded uncoloured facts* (*d*).

The ideal observer would observe *all* forms of human behaviour at a given date, *not leaving out facts* (*d*); then, without confusing his levels, and also without confusing descriptions with inferences, he would reach his higher order of abstractions properly, with very different resultant doctrines, which would produce entirely different semantic evaluation, and motivate equally different action.

We may understand now why we must constantly revise our doctrines, for the above analysis throws a considerable light on the fact that scientists need training with the Differential as much as other mortals (the author included). History shows that they have not officially checked themselves up sufficiently to become aware of this fatal habit of confusion of orders of abstractions through identification.

16. ON THE MECHANISM OF IDENTIFICATION AND VISUALIZATION

Objectification and visualization are usually not differentiated. The first represents a very undesirable semantic process, whereas the second, visualization, represents one of the most beneficial and efficient forms of human 'thought'. From a [non-A] point of view, such a lack of differentiation between the two reactions appears as a very serious problem, requiring an analysis of the respective mechanisms.

To visualize, we must have such forms of representation as lend themselves to visualization; otherwise, we must fail. The [A]-system which could not adequately handle asymmetrical relations. and could not be built explicitly on structure, necessarily involves identification. In the [A] period, we were able to visualize objects and a few objective situations, but all the higher abstractions were, in principle, inaccessible to visualization, making scientific theories needlessly difficult. A [non-A]-system, free from identification, must be based explicitly on *structure* on all levels (structure defined in terms of relations and ultimately multi-dimensional order), which can be easily visualized. It should be recalled that structure, relations, and multi-dimensional order supply us with a language which completely bridges daily-life experiences with all science, leading toward a *general theory of values*. Mathematics and mathematical physics then become the representatives and the foundation of all science; and in the human field a general theory of values will lead to adjustment or sanity and will some day include ethics, economics, .

For these reasons, the Structural Differential is uniquely useful, as, at a glance, it conveys to the eye structural differences between the world of the animal, the primitive man, and the infant, which, no matter how complex, is extremely simple in comparison with the world of the 'civilized' adult. The first involves a *one-valued orientation* which, if applied to the ∞-valued facts of life, gives extremely inadequate, wasteful, and ultimately painful adjustment, where only the few strongest survive. The second involves ∞-valued orientation, similar in structure to the actual, empirical, ∞-valued facts of life, allowing a one-to-one adjustment in evaluation with the facts in each individual case, and producing a semantic flexibility., necessary for adjustment. This flexibility is known to be the foundation for balanced semantic states, 'higher intelligence', .

Visualization requires a definite elimination, through differentiation, of harmful identification, which, as usual, is based on incorrect evaluation of structural issues. Thus, we have had endless, bitter, and futile arguments as to whether or not the 'mechanistic' point of view of the world and ourselves is legitimate, adequate, . The average person, as well as the majority of 'philosophers', identifies *'mechanistic'* with *'machinistic'*. Roughly, mechanics is a name for a science which deals with dynamic manifestations on all levels; thus, we have macroscopic classical mechanics, colloidal mechanics now being formulated, and the submicroscopic quantum mechanics already being well-developed disciplines. In the rough. 'machine' is a label applied to a man-made apparatus for the application or transformation of power. But even machines differ greatly; thus, a dynamo is entirely different in principle, in theory, and in applications from a lathe or an automobile.

If we ask: 'Is the machinistic point of view of the world justified?', the answer is simple and undeniable; namely, that this point of view is grossly inadequate and should be entirely abandoned. But it is not so with the mechanistic point of view, understood in its modern sense and including the quantum mechanics point of view, which is entirely *structural*. In 1933, we know positively that even the gross macroscopic physico-chemical characteristics of everything we are dealing with depend on the sub-microscopic *structure* (see Part X). The details are not yet fully known, but the principles are firmly established. With a [non-A] understanding and evaluation of the unique importance of structure as the only possible content of 'knowledge', these 'firmly established' principles become 'irreversibly established'. We may go further and say that the quantum mechanics point of view becomes the first structurally correct point of view and, as such, should be accepted fully in any sane orientation. If we stop identification, then we will differentiate between some simple facts. For instance, we will understand that any semantic state, reaction, or process has its corresponding sub-microscopic, structural, colloidal, and ultimately quantum mechanical processes going on in the nervous system; however, the *s.r*, or feelings of pain or pleasure., *are not* the sub-microscopic processes. These belong to different levels, but with ∞-valued semantics we can establish in principle a one-to-one correspondence between them. Thus, when we differentiate adequately, the older *machi*nistic objections disappear entirely; and, in its proper field, for structural reasons, we must preserve the *mechanistic*, and entirely abandon the too crude *machi*nistic attitudes. The *mechanistic* (1933) attitude is based on *structure* and so is indispensable for visualization; and *training in visualization automatically abolishes objectification*, which represents an important special case of all identification. From the point of view of a [non-A]-system, adjustment and sanity in humans depend, to a large extent, on their 'understanding', which is entirely structural in character; therefore, we must accept a *mechanistic* (1933) *attitude*, which, in the meantime, can be visualized.

The finding of structural means of representation facilitates *visualization*, imagining, picturing, . In the adjustment trend we start with lower nerve-impressions, 'senses', 'feelings'., lower abstractions, and these are abstracted again by the higher centres. The higher centres produce the 'very abstract' theories, which cannot be visualized for a while. The lower centres, which are involved in visualization, can deal only with structures which can be 'concretely pictured'. So we always try to invent mechanistic or geometrical theories, such as can be handled by the lower centres.

Individual 'experiences', supplied by the lower centres of different individuals, do not blend directly. They are blended in the higher centres. In them manifold experiences, whether individual or accumulated by the race (time-binding), are abstracted further, integrated, and summarized. Once this has been accomplished, structural means are sought *and discovered* to translate these higher abstractions into lower, the only ones with which the lower centres can deal. Then we can 'visualize' our theories, and the higher centres not only influence the lower centres, but the lower centres have appropriate means by which to co-operate with the higher centres in their new *non-el* quests.

The lack of explicitly structural forms of representations is responsible, also, for the difficulties which arise when the higher order abstractions are translated into the reflex-reactions of the lower centres, which can deal with 'intuitions', 'orientations', 'visualization', . The so-called 'geniuses' have a very subtle nervous system in which the translation of higher order abstractions into lower and vice versa is easily accomplished. From the point of view of forms or representations, we can have two issues: (1) we may have *el* forms of representations which are not based on structure, visualization., and cannot efficiently affect the activities of the lower centres; (2) we may have a *non-el* system based on structure, visualization., which can be translated simply, easily, and efficiently into the terms of the lower centres. These problems are of educational importance and should be worked out more fully.

In my experience with grown-ups who have had only a *short* contact with my work, I find, in many cases, that, although they may have even given their complete verbal approval of the main point of the system, yet, invariably, in practice, the full application is lacking. Obviously, the semantic importance of the present findings is not in the verbal approval alone, when that approval is not applied, but in the consistent and permanent instinctive acquisition of the new semantic attitude which involves a complete elimination of identification, allness, elementalism, .

We can teach any one to repeat verbally, by heart, instructions for operating an automobile, a piano, or a typewriter; but no one could operate them satisfactorily by reflex-action after such verbal training alone. To operate effectively and skilfully any structural complex, we must become intimately familiar with its structural working through actual reflex-training, and only then can we expect the best results. In my experience, this is true with language, and, without the *visual* Structural Differential on which we can point our finger to the objective level and urge silence., such basic semantic *reflex-training* cannot properly be given.

If we ask a man: 'Do you know how to drive a car?', and he answers 'Yes', we assume that he has acquired the proper *reflexes*. If he answers 'No, but I know *about* it', he means that he has *not* acquired the proper reflexes, but that his 'knowledge' is on purely verbal levels, non-effective in application on *non-verbal* reflex-levels. This applies fully to *s.r*; we may 'know' *about* them, but we may never apply successfully what we supposedly 'know'. To 'know' represents a multiordinal process which involves equally the activities of the lower nerve centres and of the higher. In our *el* systems we had no such distinction, and so we confused them. The older 'knowledge', when presented in *el* language, could not have been absorbed easily by the *non-el* organisms-as-a-whole. As the main task, at present, is to unlearn the older *s.r*, the new reactions need a persistent training, particularly by the grown-ups. The *non-el*, [non-A] language and method prove to have psychophysiological importance.

Although the neurological mechanism underlying identification, objectification, visualization., is not well known (1933), neurology gives us evidence that in these states, as well as in delusions and hallucinations, the actual lower nerve centres are somehow engaged. We may assume that different 'resistances', 'blockages'., in some parts of the nervous system make the passage of nervous impulses more difficult, and it seems reasonable to suppose that, in such cases, the paths travelled by the nervous currents are different.

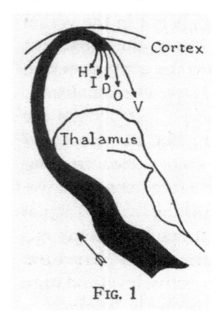

FIG. 1

In Fig. 1, an hypothetical and over-simplified scheme of the different types of distribution of nervous currents, as is known functionally is suggested. The ordering is not anatomical but functional in terms of degrees of intensity. In this scheme, we may consider that the nervous impulse (A) reaches the lower nerve centres, the brain-stem and the thalamus, passes through the sub-cortical layers and the cortex, continuously being transformed. Finally, in returning, it may take either the beneficial and adaptive semantic form of visualization (V), free from identification and semantic disturbances, or may involve identification, with semantic disturbances, such as objectifications of different orders (O), delusions (D), illusions (I), or, finally hallucinations (H).

Identification, or confusion of orders of abstractions, consists of erroneous evaluation: that which is going on inside of our skin has objective existence outside of our skins; the ascribing of external objectivity to words; the identification in value of 'memories of experiences' with experience; the identification of our *s.r* and states with words; the identification of inferences with descriptions, . Identification is greatly facilitated, if not actually induced, by the [A] structure of language in which we have *one name* for at least *four* entirely different entities. Thus, the [A] 'apple' (without subscripts and date) is used as a label for the physico-chemical process; for an object, say, 'apple1,Feb.23.1933'; for a 'mental' picture on the un-speakable semantic level, and for the verbal definition. Under such linguistic conditions, it is practically impossible, *without special training*, not to identify the *four* entirely different abstractions into *one*., with all the following sinister consequences.

Delusions represent incorrect notions and inappropriate *s.r* formed, not by insufficient knowledge or 'logic', but by affective pressure in a definite evaluational direction; as, for instance, delusions of grandeur; delusions of persecution; delusions of 'sin'; delusions of reference, .

Illusions appear more like real perceptions, but pathologically changed. For instance, anything may be semantically coloured or interpreted, or evaluated as an offense, or a threat, or a promise,.

Hallucinations consist of 'perceptions', with all their vividness, but *without* any external stimuli. Patients hear voices; see visions; feel pricks or burnings., when there is nothing to hear, or see, or to be pricked by.

In *visualization*, identification does not occur; orders of abstractions are not confused; semantic disturbances do not appear; the *evaluation is correct;* a 'picture' is evaluated as a picture and not as the events, . In other words, because of the consciousness of abstracting, the natural order of evaluation is preserved. But once, through identification, this natural order is reversed, it marks a pathological condition more or less morbid, and often of a non-adaptive character.

Identification represents, in affective tension, the mildest semantic disturbance, consisting of an error in meanings and evaluation. Objects are evaluated as events; 'ideas', or *higher* order abstractions, are evaluated as objects; as experience; as the un-speakable semantic states or reactions; otherwise, as *lower* order abstractions. The confusion in the field of higher order abstractions follows a similar rule. Inferences obviously represent higher order abstractions than descriptions; so, when they are not differentiated, higher order abstractions are again identified with the lower. We all know from daily-life experience the fantastic amount of suffering we can, and do, actually produce for ourselves and others with such identifications.

In delusions, a similar but more intense identification occurs, resulting in erroneous semantic evaluation; wishes, feelings, and other semantic states inside of our skins are projected into the external world, giving delusionally strong objective evaluation.

In illusions, we also ascribe to, or identify our complex semantic states with different perceptions and evaluate our higher order abstractions as lower.

In hallucinations, this process of reversing the natural order comes to a culminating point: higher order abstractions are translated into, and have the full vividness and 'reality' of, lower order abstractions.

We see that the pathological processes of 'mental' illnesses involve identification as a generalized symptom; which means the reversal, in different degrees, of the natural order of evaluation based on the intensified confusion of the orders of abstractions. The more intense this process of reversal becomes, the more non-adaptive and morbid the manifestations. It should be noticed that this analysis becomes a necessity once we decide to accept a *non-el* language. This analysis is far from exhaustive, but an analysis in new *non-el,* structurally correct terms, throws a new light on old problems.

Hallucinations which result from 'physical' illness do not represent a permanent danger, but when a patient seems 'physically' well, and his confusions of orders of abstractions, delusions, illusions, and hallucinations become completely 'rationalized', then these are unmistakable signs of serious 'mental' illness, suggesting sub-microscopic colloidal lesions. Now this 'rationalization' represents nothing else but a nervous disturbance and involves *identification* somewhere. In 'physical' ills the nervous system may be disturbed, but the illness does not usually originate in nervous disturbances, and so, as such, is not dangerous.

The distinction between visualization and objectification based on a [non-A]-system seems new; the difference is subtle, but when it is formulated we can discover a simple means whereby to control the situation. If we were to take a 'bone' made of papier-mache and smear it with fat or meat, Fido would, perhaps, *objectify* (identify) such a 'bone' from the smell and the form of the papier-mache with an edible one, and would fight for it. We do a similar sort of thing when we objectify. Religious wars, the 'holy inquisition', the persecution of science, which we are witnessing even at the present day in some countries and communities, are excellent examples.

We should notice that Fido was able *to trust* his natural, even 'objectified', instinct, for nature does not play such tricks on him, such as producing 'bones' of papier-mache. If nature did, dogs that objectify and persist in their liking for such 'food' would soon be eliminated. These particular objectifications would be dangerous and painful to those particular kinds of dogs with that particular nervous system, and would ultimately prove of no survival value. Thus identification, which represents an inappropriate evaluation, is harmful to all life, but is little noticed at present, because the main periods of the animal racial adjustment have been accomplished long ago. Experiments on flies show that the number of mutants which may be produced in a laboratory is large, but very few would survive outside of a laboratory. In unaided nature, these mutants probably occur, but seldom leave observable traces. However, even today, as Pavlov has shown in his laboratories, we can impose, by an interplay of a four-dimensional order of stimuli, such conditions upon animals for which their nervous survival structure was not naturally adapted, and so induce nervous pathological states. Wrong evaluation is, indeed, harmful to all life and accounts for such rigid survival laws in nature, which science teaches humans how to make more flexible. Practically word for word, this applies to ourselves. We are constantly producing more and more complex conditions of life, man-made, man-invented, and deceptive for the non-prepared. These new conditions are usually due to the application of the work of some genius, and the nervous system and *s.r of* most of us are not prepared for such eventualities. In spite of inventions and discoveries of science, which are *human* achievements, we still preserve *animalistic* systems and doctrines which shape our *s.r.* Hence, life becomes more strained and increasingly more unhappy, thereby multiplying the number of nervous break-downs.

It is known that not all people are able to visualize equally well. In the older days this fact was taken for granted, and did not suggest further analysis. Under present conditions in many human beings and also in animals, as shown in the experiments of Pavlov, the visual stimuli are physiologically weaker than the auditory ones; in man, however, the visual stimuli should be physiologically stronger than the auditory. This difference does not affect the *general* mechanism of the cyclic nerve currents and orders of abstractions. In the auditory type the main returning currents are deviated into different paths. The division between 'visual' and 'auditory' types is not sharp. In life we deal mainly with individuals who have no more than a special inclination for one or the other types of reaction.

In the case of 'mental' processes, human adjustment has to be managed on higher, more numerous, and more complex levels. Obviously, then, the auditory types are more enmeshed by words, further removed from life than the visual ones, and so cannot be equally well adjusted. This fact should not be neglected, and on the human levels we should have educational methods to train in visualization, which automatically eliminates identification.

The auditory channels which connect us with the external world are much less subtle and effective than the visual ones. The eye is not merely a 'sense-organ'. Embryology shows that the eye is a part of the brain itself and what is called the 'optic nerve' must be considered not a nerve but as a genuine nervous tract. This fact, of course, would assign to the eye a special semantic importance, not shared with other 'sense-organs' or receptors. We ought not to be surprised to find that the visual types are better adjusted to this world than the auditory types. In pathological states, such as identifications, delusions, illusions, and hallucinations, there seems to be involved a translation of *auditory* semantic stimuli into visual images. In these pathological cases the order of evaluation appears as label first and object next, while the adaptive order seems to require object first and label next. There seems little doubt that visualization is very useful, and that identification is especially harmful. The *most effective means to transform the s.r of identification is found in visualization, which indicates its special semantic importance.*

The semantic *disturbance* of identification may have many sources, auditory included, but the only adaptive trend is in visualization, which involves in some way the optical neural structure. Some structural light is thrown on this subject when we realize that, physiologically, the eye is more closely related to the vegetative nervous system, which regulates our vital organs, than the ear is. In man the optic thalamus is greatly enlarged, so that the whole thalamus is often called the 'optic thalamus'. Actually, the thalamus has many functions, other than visual, and is connected with affective manifestations.

As most of our observations are accomplished by the aid of the eye, we should expect auditory types to be *poor observers*, and so racially, in the long run, not so well adjusted semantically. Observation shows that the auditory types often have infantile reactions—a serious handicap. From an adaptive point of view the 'normal', non-infantile, best-adjusted individual ought to be a visual type. Auditory types must also be further detached from actualities than the visual types, as auditory stimuli involve more inferences than descriptions, which is the opposite of the functioning of the visual types. If inferences, rather than descriptions, are involved, we naturally deal with higher abstractions first, and with the lower next, and so there is always a danger of the semantic confusion of orders of abstractions, which necessarily involves inappropriate evaluation, of which objectification is only a particular case.

Even to common sense it seems clear that there is a significant difference between 'knowing' this world by hearing and 'knowing' it by seeing. There is, likewise, a difference between the translation of higher abstractions into lower terms by the visual path, and the corresponding translation by the auditory path. In daily life we never say 'I hear' when we wish to convey that we understand; but we say 'I see'. When we say 'I hear', we usually wish to convey that we have heard something which we did not fully grasp or approve. The above relation is rather important, but has not been sufficiently analysed. The problems of introversion and extroversion are connected with it.

There is, however, one point that I wish to make entirely clear. From the older point of view, one might say that a [non-A]-system may lead to 'over-rationalization' and, consequently, take 'all the joy out of life'. Such objections are entirely unjustified. First of all, the [A]-system leads to shallow, but often clever verbal interplay of definitions, mostly non-similar in structure to the world and ourselves, representing a species of apologetics, and usually called 'rationalization'. The [non-A]-system leads to structural adjustment of language and s.r, and a *structural* enquiry, resulting in *understanding*. It makes shallow infantile 'rationalization', 'wishful thinking', and apologetics of different brands impossible, but leads to a higher order of adult intelligence, based on *proper evaluation*. In mere 'rationalization', we often have clever, but shallow, infantile evaluation, based on the ignorance or disregard of structural facts, which alone make up the content of all 'knowledge'. In a [non-A]-system, by eliminating the sources of infantile evaluation and reactions, we supply the nervous system of the infant with uniquely appropriate material, so that it may develop into a 'normal' adult. In the older system, instead of helping, we hindered the development of adult standards of evaluation, with well-known results. There is nothing wrong with 'human nature' or the majority of nervous systems as such, but there is something definitely wrong with our educational methods inside and outside our schools.

17. ON NON-ARISTOTELIAN TRAINING

For reasons already explained, students should not only hear and see the explanations, but should also *perform for themselves*, should handle the labels and indicate with their hands the different orders of abstractions. After preliminary explanations, the children should be called to the Differential, and, using their hands, they should explain it. This applies, also, to grown-ups and to patients. The Differential is not only a permanent structural and semantic reminder which affects many nerve centres; it is more, for, in training, it conveys the *natural order* through all centres. Any reader who refuses to use his hands in this connection handicaps himself seriously, because *ordering* abolishes identification.

Fundamentally, there is no structural difference between the use of language and the use of any other mechanical device; they all involve reflex-action. It is well known that any pianist, telegraph operator, typist, or chauffeur would not be a successful performer if he had to meditate about every move he makes. As a rule, *verbal* explanations of the working of the respective machines are necessary at first, yet the structural reflex-skill required is actually acquired by prolonged practice, where again all nervous centres are involved. We all know what amazing *unconscious* reflex-adjustments a good driver of a car can make in case of unexpected danger.

A similar semantic reflex-skill is required in handling our linguistic apparatus, and, in case of danger, of sudden turns and twists, our orientation should also work unconsciously. That is why the structural *feeling* for the working of the apparatus is required. All nerve centres should be trained to employ the most effective means to affect the organism and its working as-a-whole.

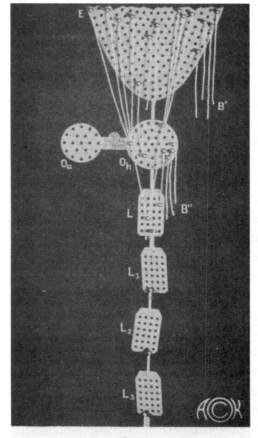

Fig. 3
THE STRUCTURAL DIFFERENTIAL

The semantic training of grown-ups and that of children do not differ in essentials. Children have fewer established habits, have more fluid *s.r* than adults, and, therefore, the results with children are achieved more quickly and last better.

I shall now explain how to train children. A similar method applies to adults, also; but an adult should not trust himself too much that he has completely acquired 'consciousness of abstracting'. He must train very thoroughly. I speak from personal experience. Although I have the Differential before my eyes practically always and am the author of the present system, yet every once in a while I catch myself with one of the old vicious semantic habits. Habits, and particularly linguistic habits, may be very pernicious and difficult to change.

We do not need to start with profound theoretical considerations; we may start with any familiar daily-life objects and a microscope or magnifying glass. We bring the Differential into the classroom, with labels (except one) detached, but do *not* proceed to explain it. We start with a little semantic *experiment* upon the subject of 'allness'. We take any actual object, an apple, a pencil, or anything else which is familiar to the children. The principles involved are entirely general and apply to all objective levels in a very similar way. We tell them that we will have some fun. Then we ask them to tell us 'everything' or 'all' about the object in question; in this case, the apple. When the children begin to tell us 'all' about it, we write the characteristics down on the black-board. *This last is vital*. We must have a visual and extensional record of the ascribed characteristics. When the children have exhausted their ingenuity in telling 'all' about the apple, we should *not* be satisfied. We should make them doubt, urging them that, perhaps, they did not tell 'all' about it, using the word 'all' continually. The term 'all' should be stressed and repeated to the point of the children's being thoroughly annoyed with the term. The more they learn to dislike this term, the better. We are already training a most important s.r.

We should *not* be satisfied with the best answers made by the most intelligent children. In a large class there may even be a child who tells us bluntly that it is impossible to tell 'all' about the apple. We should concentrate on the *less* intelligent children and deal particularly with them. There are many and important reasons for this. For one thing, the children become more eager and more interested in their own achievement. Then, too, they easily learn by example what a difference in intelligence means. This understanding of the shortcomings of others has an important semantic, broadening effect. In life, numerous serious 'hurts' occur precisely because we do not appreciate some natural shortcomings and expect *too much*. Expecting too much leads to very harmful semantic shocks, disappointments, suspicions, fears, hopelessness, helplessness, pessimism, .

The less bright children benefit also. The experiment is conducted on their level, so that they also have the maximum chance to benefit. Soon the children begin to argue about the new method and to explain it to each other by themselves; for we have touched very vital and complex semantic processes of 'curiosity', 'achievement', 'ambitions'., characteristics strongly represented in the child's life. We evade, also, the danger of taking clever, yet shallow, replies as standard. The last error would be fatal, as the issues are fundamental, and we should not rest content with mere verbal brilliancy.

When the subject seems exhausted, and the list of characteristics of the apple 'complete' (we repeatedly make certain that the children assume they have told us 'all' about it), we cut the apple into pieces and show the children experimentally, using eventually a microscope or a magnifying glass, that they did *not* tell us 'all' about the apple.

It may appear to some educators that such training might involve some undesirable psychological results. But later, when consciousness of abstracting is acquired as a lasting semantic state, this fear appears entirely unjustified, as explained further on. The first step in dealing with 'reality' seems to demand that we abandon entirely the older delusional methods.

When the children have become thoroughly convinced of the non-allness and the *impossibility* of 'allness', we are ready to explain to them what the word abstracting means, using again the terms 'all' and 'not all'. We show them a small rotating fan and explain to them about the separate blades which when rotating we see as a disk. In such demonstrations we can go as far as desired. All science supplies data (e.g., the dynamic structure of seemingly solid materials). We must select the data according to the age of the children or the knowledge of the grown-ups. Everything said should be demonstrated empirically from a structural point of view.

The next step is to demonstrate practically that an object taken from different points of view has different aspects for different observers. We may use different objects or wooden geometrical figures painted with different colours on different sides. We may place the object in different positions and ask the children their descriptions, which should be written down. The descriptions will, of course, be different, and the children should be made thoroughly aware of this. In all these preliminary exercises the ingenuity of the teacher has a vast field for exercise, and we do not need to enter into details.

When all these results have been accomplished on the level of the *least developed* child, we then proceed to explain the Differential as a structural diagrammatic summary of the above results. It is a *positive condition* that the new language be used, and that an object be described as an *abstraction* of some order. If this vital structural point is disregarded, most of the psycho-logical semantic benefits of 'non-allness' are either lost or greatly lessened. We should make this term clear to the child, and should train him in its use, as it appears uniquely in accordance with the structure and the functioning of his nervous system. The child should be warned that the old languages are not structurally suitable for their future understanding and semantic adjustment. This warning should be repeated seriously and persistently.

Having eliminated 'allness', we begin to eliminate the 'is' of identity, which, at the primitive and infantile stages of racial human development, happens to be extremely ingrained in our *s.r*, embodied, as it is, in the structure of our daily language. As was explained before, identification is a natural reaction of the animal, the primitive man, and the infant, reflected and systematized in the [A] and older linguistic systems, which, through the ignorance or neglect of parents and teachers, is not counteracted and so is continued into the lives of children and grownups, until, finally, it becomes embodied in the structure of what we call 'civilization' (1933). In a theory of adjustment or sanity we must counteract this animalistic, primitive, or infantile *s.r* by building a [non-A]-system, which entirely rejects the 'is' of identity.

In the [A]-system, through the use of this 'is', different orders of abstractions were unconsciously identified in values, in obvious contradiction to empirical facts. In other words, being identified in values, they were treated as of one order or on one level and so did not necessitate indefinitely many expanded orders of horizontal and vertical differences. Similarly with the objectively meaningless 'infinite velocity' of a process, it does *not* allow *order*. But once we have a finite velocity of a process, *order* makes its appearance as an indispensable aspect of a process. *The finite and known velocity of nerve currents on the physico-mathematical levels results in ordered series on physiological levels; in non-identity and proper evaluation on semantic levels, and in orders of abstractions and a non-aristotelian system and general semantics on verbal levels.*

Once we abolish in our language the always false to fact 'is' of identity, we automatically stop identifying different orders of abstractions. We do not assume that they represent one level, which becomes expanded into a natural ordered series of indefinitely many different orders of abstractions, with different values. Adjustment, therefore, sanity and adulthood of humanity, depend on proper evaluation, impossible under conditions of delusional identification of fundamentally different orders of abstractions. We must then train the *s.r* in the natural *physiological order* of the process of abstracting which, on the psychological levels, become non-pathological semantic evaluation.

In the case of training in the 'non-allness', it was necessary to start with the analysis of an ordinary object, to give the child a simplified theoretical explanation, and then to demonstrate it empirically. The child will be easily 'convinced', but this conviction is not enough, because it will not affect permanently his *s.r.* We explain this difficulty very simply, telling him that, although he 'agreed' with our presentation, he will very soon 'forget' it, and so we need a permanent visual reminder which is supplied by the strings, freely hanging from the event and from the object, and indicating those 'characteristics left out', or not abstracted.

In the elimination of the 'is' of identity, we have also structurally interconnected aspects. The rejection of this 'is' becomes an equivalent to the stressing of the *stratification* in the structure of 'human knowledge'. To facilitate training, we should *stress both aspects* by all available means, and should involve as many nerve centres as we can. Thus, through the ear we stress verbally the formula of the rejection of the 'is' of identity by indicating with our finger the different orders of abstractions, in the meantime, affecting the eye while we repeat 'this *is not* this'. We utilize the kinesthetic centres, not only by pointing the finger to different levels, but also by making broad motions with our hands, indicating the stratifications. We should train in both horizontal and vertical stratifications, always using the hands. The horizontal stratification indicates the difference, or ordering of different order abstractions; the vertical stratification indicates the difference between 'man' and 'animal' and the differences between the different absolute individuals. In both cases, the semantic effect of the 'is' of identity is counteracted.

The above procedure in training has an important neurological foundation. Besides what has been explained already, we find that a word has four principal characteristics with corresponding cortical representations. A word can be heard, seen, spoken, and written. Language, thus, involves many nervous functions; e.g., auditory, visual, and diversified motor nerve centres, interconnected in a most complex network of 'horizontal' and 'vertical' fibres. The use of the Differential involves all available nervous channels; we see, we hear, we speak, we move our hands, indicating stratification, 'non-allness'., engaging large cortical areas, and so have the maximum probability of affecting, through *non-el* methods, the organism-as-a-whole. The Differential gives us a special, simplified, yet advanced interracial *structural symbolism* (1933), which affects wide nervous areas of the illiterate, or nearly illiterate person, or of the infant., which otherwise could not be affected. It is known that extensive reading and writing, as well as speaking a number of languages, has a very marked cultural effect and helps visualization and consciousness of abstracting. The reason can be found, perhaps, in the fact that a learned polyglot, or a scholar, utilizes many nerve centres in co-ordination. In the older days, unless one became a scholar of some sort, it was extremely difficult to train these nerve centres in co-ordination. With the Differential we can train simply, and comparatively quickly, all necessary nerve centres, and so impart to children and to practically illiterate persons the cultural results of prolonged and difficult university training without any complicated technique. This last should always be regarded as a means and not as an end.

In my experience with children, and with men from the lowest 'mentality' to the highest, the non-identity of different orders of abstractions is usually taken lightly. It all seems so simple and self-evident that no one assumes that there could be serious, unconscious, structural, semantic, linguistic, and neurological delusional mechanisms involved, which cannot be reached without specially devised non-identity training. The delusional feelings of 'allness' and 'identity' are peculiar in that, like other pathological states, they tend to appear as all-pervading. It is the most difficult in daily, as well as in medical, experience to make a breach in this all-pervading tendency, but once this delusional state is even partially replaced by glimpses of *m.o* reality, the further elaboration and training in adjustment to 'reality' becomes comparatively simple. Thus, in practice, if we *start* with ordinary objects, feelings, and words, and train in the non-allness and non-identity, any child, or any grown-up, even an imbecile, can follow this easily. Once this feeling has been acquired, and in most cases it is only a matter of method and persistence to acquire it, the main *semantic* blockage has been eliminated, and the rest is comparatively easy. I have had no opportunity yet to verify it, but I am convinced that even a superior imbecile could be trained to differentiate between descriptions and inferences, after he has learned to differentiate between the objective levels and words. In such a training with imbeciles we can go in simplicity as far as desired; thus, if the given individual is hungry and says he wants 'bread', we hand him a label which is attached to the objective bread, and he would be quickly made to realize that the symbol *is not* the thing symbolized.

It should be realized that in the training we should impart the obvious fact that words or labels represent conveniences, and *are not* the objects or feelings themselves. We should carry the labels in our pockets, so to say, as we carry our money, or checks for hats or trunks, and not identify them 'emotionally' with what they eventually stand for, because monetary standards change, and hats and trunks get exchanged, lost or burnt. To accomplish this, we must have *objective labels*, which we may handle and carry in our pockets, and also an objective something to which we can attach the labels. In the present [non-A]-system the rejection of the 'is' of identity is complete and applies to all levels. Thus, the event *is not* the object; the object *is not* the label; description *is not* inference; a proper name *is not* a class name.; the characteristics ascribed to events, objects, or labels *are not* identical, an object, a situation, or a feeling *is not* identical with another object, situation or feeling., ., all of which establishes a *structure of horizontal and vertical stratification*. At an early stage of the training, we must begin with what appears the simplest and most obvious to the child; namely, the absence of identity between the word and the object, or that the word *is not* the object. We accomplish this by stressing that one cannot sit on the *word* 'chair', that one cannot write with the *word* 'pencil', or drink the *word* 'milk', . These simple facts should always be translated into the *generalized form*, indicating with the hand the two levels on the Differential, conjointly with the fundamental formula 'this *is not* this'. We should always tell the child that the formula is entirely general, but for the present we should not go into any further details.

At this stage we can advance one step further, still using *only* ordinary objects as examples, and explain the un-speakable character of the object; namely, that whatever we can see, taste, smell, handle., is an absolute individual (demonstrated empirically) and *un-speakable*. We then take the apple, bite it (actually performing), and explain that, although the object is *not words*., yet we are very much interested, and traditionally so, in this un-speakable level. Then we explain repeatedly and at length, emphasizing the important principle of evaluation, that to live we must deal with the objective level; yet this level cannot be reached by *words alone*. As a rule, it takes a few weeks, or even months, before this simple s.r is established, the old identification being psychophysiologically very much ingrained. Once this is established, we stress the fact that we must handle, look, and listen., never speak, but remain silent, outwardly as well as inwardly, in order to find ourselves on the objective level. Here we come to one of the most difficult steps in the whole training. This 'silence on the objective level' involves checking upon neutral grounds of a great many 'emotions', 'preconceived ideas', . This step, in the meantime, appears as the first, the simplest, most obvious and most effective psychophysiological 'reality-factor' in eliminating the delusional identifications.

Once the child is thoroughly aware of the absence of identity between words and objects, we may attempt the expanding of the notion 'object' to the 'objective levels'. Such training requires persistence, even though it seems fundamentally simple. We demonstrate and explain that action, actual bodily performance, and all objective happenings, *are not* words. At a later stage we explain that a toothache, or demonstrate that the actual pain of a prick., *are not* words, and belong to the *objective un-speakable levels*. Still later, we enlarge this notion to cover all ordinary objects, all actions, functions, performances, processes going on outside our skin, and also all immediate feelings, 'emotions', 'moods'., going on inside our skins which also *are not* words. We enlarge the 'silence' to all happenings on the objective levels and the animalistic, 'human nature' begins to be 'changed' into quite a different *human* nature.

When this is accomplished the rest is much simplified, although much more subtle. We explain, as simply as we can, the problems of evaluation and *s.r*, stressing and making obvious the fact that our actual lives are lived *entirely* on objective, un-speakable levels. We illustrate this all the time by simple examples, such as our sleeping, or eating, any activities, or pain, or pleasure, or immediate feelings, 'emotions'., which *are not* words. If words are not translated into the first order un-speakable effects, with the result that we do not do something, or do not feel something, or do not learn or remember something., such words take *no effect* and become useless noises.

One fact should be stressed; namely, that the problem is *not* one of 'inadequacy of words'. We can always invent 'adequate words', but even the most ideal and structurally adequate language will *not be* the things or feelings themselves. On this point there is *no possible compromise*. Many people still utter quite happily, pessimistic expressions about the present *language*, based on silent assumptions connected with unconscious delusional identification, and believe that in an 'adequate' language the word by some good primitive magic would be identical with the thing. The more the denial of the 'is' of identity is driven home, and the sooner it becomes a part of one's *s.r*, the sooner the 'consciousness of abstracting' is acquired.

We are now ready to go further into the theory of *natural evaluation* based on natural *order*. As a preliminary step, we must show repeatedly the difference between descriptions and inferences, using simple examples. We must stress the fact that words, as such, must be divided into two categories: a first, of descriptive, in the main, functional words; and a second, of inferential words, which involve assumptions or inferences. Thus, 'A does not get up in the morning' may be considered as descriptive. If A explicitly refuses to get up, the statement 'A *refuses* to get up in the morning' may also be considered as descriptive. If A did not explicitly refuse, this statement becomes inferential, because A may be dead or paralysed. If we would say simply, 'A is lazy', such a statement represents an illegitimate inference of high orders based on ignorance, because in 1933 it is known that 'laziness' represents a symptom of physico-chemical, colloidal, or semantic disturbances. It should be stressed that this discrimination between descriptive and inferential words, although extremely important, is not based on any 'absolute' differences, but, to a large extent, depends on the context. I shall not analyse this problem further, because any parent or teacher who has acquired the consciousness of abstracting himself will find more examples at hand than are needed.

18. ON INFANTILISM IN ADULTS)

As has already been mentioned, the main symptoms of physical and 'mental' illnesses are few and simple. This would suggest the possibility of simple and more general theories relating to the fundamental symptoms. The colloidal structure of protoplasm accounts for this peculiar simplicity and for the small number of the fundamental symptoms. In the 'mental' field these fundamental symptoms are accounted for by a simple structural, functional principle of 'copying animals' in our nervous processes, which must be harmful, and which is characterized by the lack of consciousness of abstracting, implying colloidal disturbances. Psychogalvanic experiments show clearly that every 'emotion' or 'thought' is always connected with some electrical currents, and that electricity seems fundamental for colloidal behaviour, and, therefore, for physical symptoms and the behaviour of the organism.

In the colloidal processes we find the bridge between the 'physical' and the 'mental', and the mutual link seems mainly electricity. It is more than mere coincidence that all illnesses, 'physical' or 'mental', have only a few fundamental symptoms; and we should no longer be surprised to find that physical ills result in 'mental' symptoms, and that 'mental' ills may also involve 'physical' symptoms.

If a simple symptom is completely *general,* it indicates that it is structurally fundamental, and we shall be repaid if we devote special attention to it. As a rule, in 'mental' ills we observe a striking appearance of symptoms which have a sinister parallel with the behaviour of infants. Arrested development or regression in grown-ups also exhibits these infantile characteristics. In other words, whenever infantile characteristics appear in grown-ups it indicates that the 'adult' has not grown up fully in some semantic respects, or has already started on the way of regression, implying some colloidal or *m.o* structural injury.

When we speak of 'infantilism' in 'adults', we include symptoms which belong to the period of childhood in its organ erotic or autoerotic and narcissistic stages. It should be recalled that in children these semantic phases are natural; they become pathological only when the individual does not outgrow them and exhibits them as a grown-up. The term 'infantilism' is a rather sinister one, and should never be applied to children. Children behave like children, and that ends the subject. But children have fewer responsibilities, their sex impulses are undeveloped., and so their behaviour cannot be equally dangerous to themselves and others. But not so with grown-ups. They have responsibilities, duties, often strong sex impulses., which make out of the infantile 'adult' an individual dangerous to himself and others. The term 'social' period, or 'socialized' individual, is sometimes wrongly interpreted. The fact that human achievements and capacities are accumulative and depend on achievements of others makes us, by necessity, a social time-binding class of life, which again involves more complex modes of adjustment. Whether we approve or disapprove the existing legal and police regulations has nothing to do with the fact that in a social class of life some restrictions are necessary. Our present commercial 'civilization' can be characterized as of an infantile type, governed mostly by structurally primitive mythologies and language very often involving primitive *s.r.* One need but read the speeches of different merchants, presidents, and kings to be thoroughly convinced of this. The rules and regulations are naturally antiquated, and belong to the period to which the underlying metaphysics and language belong. The 'adult' or scientific semantic stage of civilization would be precisely the 'social' stage of complete evaluation of our privileges and *duties.*

In speaking about infantilism, it should be remembered that the child has an advantage over the imbeciles, idiots, and 'mentally' ill who have stopped development or have regressed to the age of the infant or the child. The 'normal' child profits by experience and outgrows the semantic characteristics that are natural to a given age. In cases of arrested development or regression, the undesirable infantile characteristics persist in the grown-ups and are a source of endless difficulties and suffering to them and their associates. Thus, in our childhood we all have had experiences similar to that of the patient of Dr. Jelliffe (see *Science and Sanity*, p. 496-498), and are no worse off because of them. But, if the reader should imagine himself in the position of the patient with those infantile characteristics, he would realize that an enormous amount of suffering, fear, shame, bewilderment., results for the patient. The worst feature in such cases is found in the fact that an infantile type usually cannot 'outgrow' or alter such characteristics by himself, and needs very wise and patient outside help in re-training, or medical assistance, if he is ever to overcome earlier inappropriate *s.r.* But if we *start* the education of an infant with appropriate *s.r,* such a procedure must play a most important preventive evaluational role.

Before birth, the child can be considered as in ideal conditions. He floats comfortably in a fluid of a temperature equal to his own. All his wants are satisfied, as everything is supplied to him by the maternal body. At birth, the child must begin to breathe, and a little later he must take food, digest, . External influences begin to impinge on him, and he must begin to adjust himself. Very soon the average infant finds that he can get what he wants, within certain limits, by certain movements or by crying. For the infant, a cry or a word becomes semantic magic. In Pavlov's language, a word governs a conditional reflex. In psychiatry, a definite series of such conditional *s.r* of animalistic low order of conditionality is called a 'complex'. In Pavlov's experiments a dog is shown food and a bell rung simultaneously. At the sight of food, saliva and gastric juice flow. Associations soon *relate* the ringing of the bell and the food, and, later, simply the ringing of the bell will produce the flow. In another animal some other signal, a whistle, for instance, would produce similar effects. In different people, through experience, associations, relations, meanings, and *s.r* are built around some symbol. Obviously, in grown-up humans the identification of the symbol with the thing must be pathological. But in infancy the confusion of orders of abstractions must be considered as an entirely natural semantic period. The infant 'knows' nothing about science and events. Objects and 'sense perceptions' 'are' the only 'reality' he knows and cares about; so he does not and cannot discriminate between events and objects. By necessity, he identifies unknowingly two entirely different levels. As his symbol usually means a satisfaction of his wants, naturally he identifies the symbols with the objects and events. At this stage, also, he cannot know that the orders of his abstractions can be extended indefinitely, or that his most important terms have the multiordinal character. It is important to notice that objectification, and, in general, identification or confusion of orders of abstractions, are semantically *natural* for the infant. The more the child comes in touch with 'reality', the more he learns, and in a 'normal' child the 'pleasure principle', which was established as a method of adjustment on the infantile level, is slowly displaced by the 'reality principle', which then becomes the semantic method of adjustment of the complete adult. Science alone gives us full knowledge of current 'reality'. But science represents a social achievement, and, therefore, a complete adult, in growing up to the social level, must become aware of the latest stages of *m.o* reality. These are given by the current scientific methods and structural notions about this world, and gradually become incorporated in the structure of the language we use, always deeply affecting our *s.r.*

It is important that in the twentieth century we should realize that the work of Einstein and the four-dimensional space-time continuum establishes a language of different structure, closer to the facts we know in 1933, and that it gives us a new semantic method of adjustment to a new 'reality' (see Part IX).

The semantic stages of the development of the child must naturally pass through the stages outlined above. When he begins to differentiate himself from the environment, he is self-centred and concentrated on his 'sensations' (autoerotic). Later he projects his own sensations on the outside events; he *personifies*. This semantic trait is often found in incomplete adults, when in anger they break dishes or furniture.

The child is interested, first, in himself (autoerotic); then in children like himself (homosexual). Slowly his interests turn away to persons less similar to himself, to the opposite sex, and so he enters the semantic period of the race development.

Similar semantic processes are to be seen in the racial developments as given by anthropology, and are reflected in the structure of the languages. In the archaic period of one-valued 'pre-logical thinking', which is found among primitive peoples, the 'consciousness of abstracting' is practically nil. The effect produced by something upon an individual inside his skin is projected outside his skin, thus acquiring a demonic semantic character. The 'idea' of an action or object is identified with the action or the object itself. Identification and confusion of orders of abstractions have full sway.

The paralogical stage is a little more advanced. In it the identification is based on *similarities*, and differences are neglected (not consciously, of course). Levy-Bruhl describes this primitive semantic period by formulating the 'law of participation', by which all things which have *similar* characteristics 'are the same'. A primitive syllogism runs somewhat as follows: 'Certain Indians run fast, stags run fast; therefore, some Indians *are* stags'. This semantic process was entirely natural at an early stage and laid a foundation for the *building of language* and higher order abstractions. We proceeded by similarities, much too often considered as *identities*, with the result that differences were neglected. But in actual life, without some primitive metaphysics, we do not find identities, and differences become as important as similarities. The former primitive emphasis on identity, later enlarged to similarities, must, at some stage of human development, become semantically disastrous and the optimum adjustment an impossibility.

In building a [non-A]-system, we have to stress *differences*, build a 'non-system' on 'non-allness', and reject identity. The older semantic inclinations and infantile or primitive tendencies were a necessary step in human evolution. For sanity, we must outgrow these infantile semantic fixations. Similarly, for civilization, we must grow out from primitive structural fixations, primitive metaphysics, taboos, and other primitive s.r. These primitive habits, languages, and structural metaphysics and reactions have been extremely ingrained in us through the ages, and it requires effort and new semantic *training* to overcome them.

In the 'mentally' ill we find sinister and very close parallels to the behaviour of the primitive man and the infant, not only in the 'mental' and 'emotional' responses, but even in physical behaviour, postures, drawings, and other modes of expression. These parallels are today recognized by practically all scientific workers and are analysed in many excellent volumes.

We should notice that in this maze of observational material, one general rule holds; namely, 'consciousness of abstracting' offers a *full semantic solution*. In it we find not only a complete foundation for a theory of sanity, but also the semantic, psychophysiological mechanism for the passing from the infantile, or primitive-man, level to the higher level of complete adulthood and civilized social man.

In the field of higher abstractions the train of 'ideas' of children, imbeciles, and idiots is restricted. Uncommon 'ideas' are left out, and only those which originated in immediate 'sense perception' are easily grasped. Until lately, even in science, such an attitude was noticeable, as for instance, in gross empiricism, or in the case of the physicist already mentioned who was willing to 'fight' to prove that he 'saw' the 'electron', . He did not realize that inferential entities are just as good abstractions as those he 'sees'. The attitude of the 'practical man' who pooh-poohs science and the 'highbrows' may serve also as an example.

Children, idiots, and imbeciles cannot comprehend anything complicated; they see some elements, but miss the relative wholes. We have elaborated a racial language of 'senses' and elementalism. Similarly, in schizophrenics the relative whole is disregarded, while, on the other hand, a single semantically effective characteristic is sufficient to connect the most heterogeneous abstractions in an unnatural whole. Word-relations have a predominance over actualities (identification). Thus, a patient looks anxiously at a moving door and exclaims 'Da fressen mich die Thuren' and refuses to pass through the door-way. [Animal = Thier, Door = Thur, so that the unconscious play upon words gives the meaning: 'Doors devour me', for 'Animals devour me'.] Here we see the identification of words with objects carried to the limit. In general, the *s.r* of the schizophrenic seem such that he identifies intensely his higher abstractions with the lower.

Much excellent material on infantilism can be found in Dr. Joseph Collins' *The Doctor Looks at Love and Life*, particularly in his chapter on Adult Infantilism, from which much of the following material is taken. and which I acknowledge gratefully.

Children and idiots live in the present only and do not concern themselves with the past and the future beyond their immediate gratification. Infantile types also want all the 'sense' enjoyment of the moment, never enquiring about the sufferings of others or of the consequences for themselves in the future. Indeed, their attitude is often hostile toward those who take into consideration a larger field. 'Aprés nous le deluge' represents their royal semantic motto. On national and commercial grounds, they devastate their natural resources, since they are interested only in some immediate and selfish advantage. They love praise and hate blame, not realizing that a *critical* attitude gives the foundation for proper evaluation and becomes a semantic characteristic of maturity and that, generally, it is *more beneficial* in the *long run*. They thrive and thrill on commendations and compliments, and shiver and shrink at disapproval. Such characteristics are found even in whole nations. They are self-satisfied, and keep aloof from others in international affairs, not realizing that this is impossible, and that the attempt is ultimately harmful to them. They assume, as an excuse, the superiority of their institutions., and the 'righteousness' of their own conduct.

Children and superior idiots appreciate resemblances more readily than differences. Simple generalizations are possible, but often they are hasty and faulty. A child's pride and self-respect are hurt if he is considered different from other children, or is dressed differently. Originality and individuality are tabooed among children. Because of semantic undevelopment, differences become a disturbing factor to them; they want everything standardized. On national grounds, the adult infants standardize all they can and have even a kind of hostility to anything which has an individual flavour. For instance, those who wear straw hats after an arbitrary date are attacked on the streets. Not wanting to 'think', or to bother about differences, they fancy that they can regulate life by legislation and they keep busy manufacturing 'laws', which are very often impracticable and self-contradictory. When they pass several thousand 'laws a year, these become a maze and a joke. The ultimate semantic result of such over-legislation is a complete lack of justice or of any respect for 'law'. Not being able to 'think' for themselves, they leave that bothersome function to politicians, priests, newspapermen, . Under such conditions life is impossible without expensive lawyers.

Not having the critical semantic capacity for proper evaluation, their likes and dislikes are very intense. They cannot differentiate the essential from the unimportant. The immediate 'sense' perception or 'emotion' unduly influences their actions. Impulses to copy others dominate them. They are often prejudiced. This results in weak judgement, over-suggestiveness, 'emotional' outbreaks, exaggerated sensibility, variability of affective states., and, finally, in an attitude toward life devoid of proper evaluation. Their moods are changeable; their attention readily gained and as readily diverted. They become easily intimidated and frightened, and easily influenced by others.

The above semantic characteristics are sponsored by commercialism, and build up the kind of methods, advertisements, and business policies which we see about us. This also introduces a semantic factor of disintegration into human relationships, as it leads to methods of trickery, to 'putting something over' on the other fellow, and appeals to self-indulgence, . When such commercial tactics are national, their sinister educational effect is pronounced. Children, from the age when they begin to read, are impressed by such practices as *normal* and take them as semantic standards for their own further orientations. Unfortunately, even psychiatrists have not, as yet, analysed the semantic influence of such advertisements on the building and preserving of infantile characteristics.

Children lack moderation and a semantic sense of proper evaluation. Tolerance is not one of their characteristics. To them persons and 'ideas' are evaluated in extremes, either good, 'wonderful', or bad, 'terrible'. Their *s.r* appear dogmatic and stubborn, as in all the unexperienced. They talk too much or are silent; they praise too much or blame too much; they work too hard or play too hard, and know no middle ground. The whole life of a nation may be coloured by such semantic attitudes. Nations become boastful of their own possessions and achievements, and happily borrow and forget the achievements of others. They pride themselves on having the largest airships, the largest cities, the highest buildings, the longest bridges, . They know no moderation in food or drink; they eat or drink too much or become total 'prohibitionists'. They exhibit quick friendships and quick dislikes. They are solemn in their games, like children who are playing father and mother, and make out of games a national event. The childish pleasure of defeating an adversary accounts for national crazes, like racing, boxing, football, baseball, and similar sports, which often overshadow in public attention all really important issues.

Children and many idiots are incapable of any choice which involves meanings and evaluation. When confronted with a situation in which they have to choose between two alternatives, they have difficulties, and often want both. Similarly with 'ideas'; they often keep sets of entirely self-contradictory 'ideas'. Even scientists of an infantile type do so, and then publish 'manifestos' in which they try to justify such behaviour and semantic attitudes. Merchants train salesmen especially to induce customers with such infantile *s.r* to buy what they do not need. This attitude is often extended to marriage. Any man and woman may marry simply because they come across each other; then, when they meet somebody else, they soon change the object of their sentiments.

All classes of feeble-minded and children show marked credulity; they like fairy tales and fantastic stories. Free inventions, by a process of objectification, are taken as experience. Children and schizophrenics *pun and play on words*. They build up languages of their own. Perseveration and stereotypy in speech are also found among them. National commercialism utilizes this principle in advertisements and tries to run a country by verbal slogans and play on words.

Psychopathology and experience show that the 'self-love', 'self-sufficiency'., of infantilism are usually accompanied by marked sex disturbances, which, from a racial point of view, are just as important as the semantic disturbances.

Infantile types often have 'charming' qualities. The women 'are sweet', 'nice'; the men seem 'good mixers' and 'popular'. The opposite sex often likes these characteristics. In men a feeling of sympathy is aroused for the 'helpless little girl', or else pedophilic tendencies are released. (Pedophilia is used as the name of a 'mental' disturbance or desire for relations with children, often found among senile dementia and imbeciles.) In women, often a feeling of motherhood leads them to like infantile males. The charm of the child lies, to a great extent, in his narcissism, his self-sufficiency, and inaccessibility. Certain animals, such as cats and larger beasts of prey, fascinate us, as they do not concern themselves with us and are inaccessible. But this 'charm' has another and very tragic side. Such infantile types cannot stand responsibilities; their affections are shallow and unreliable; they know how to take, but do not know how to give, . In life, such connections lead invariably to great unhappiness, and often to disasters. The children produced by such infantile types are usually completely ruined by the lack of parental understanding or lack of care. Instead of liking such types, men and women of semantic maturity should either avoid them or suggest psychiatrical consultation.

Infantile types invariably show some sex disturbances, which also add greatly to family and social difficulties. The men are often impotent; the women, frigid. Onanistic and homosexual habits or tendencies persist, although such an adult infant is married and has an opportunity for normal life. A very important [non-A], *non-el* fact should be noticed. Since the organism works as-a-whole, 'mental' components should be considered in connection with sex life. An infantile type appears still in an organ erotic stage. He wants only sense gratification. From a theory of sanity point of view, *prostitution appears as a substitute for onanism.* In adult infants, we very often find either impotence, frigidity, onanism, homosexualism, as simple forms of arrested development or regression, or more extreme forms, like many cases of prostitution. Infantiles not only indulge in promiscuity, but build up fanciful rationalizations and represent their own infantile tendencies by 'theories' as 'normal' conduct. Many criminals, professional 'vamps', and professed 'heartbreakers' belong to this type. It is interesting to note that many 'mental' illnesses are connected with different onanistic rationalizations. Often excessive cleanliness, continual washing of hands., appears. If a schizophrenic onanist has a melancholic make-up, he rationalizes his problems that he is 'rotting because of his sins'. If he has a manic make-up, he feels that he 'is a saviour of mankind'.

In all such cases family life is very unhappy, and the future of the children bred under such conditions is usually gloomy. Children need healthy family and semantic conditions to develop into healthy individuals.

The majority of professional criminals and prostitutes have an infantile make-up. No matter how cunning, they usually show little foresight. They appear egotistic, boastful, exhibitionistic, . Gangsters love pomp; their funerals are, as a rule, very expensive,—they want, even after death, to 'show off'. Criminals seldom become good fathers or mothers. They treat each other brutally and are generally promiscuous. Ethically, they behave usually as 'moral imbeciles', not realizing fully what they do. I am not advocating the abolishment of the death penalty on sentimental grounds, but an *enlightened* society should abolish any *penalty* on sick individuals. The 'mentally' ill criminal type should be either taken care of or else eliminated with some scientific benefit, but *not* as a *penalty*. Professional criminals can hardly ever become 'morally reformed' or useful members of society, unless the application of medical science can alter their pathological *s.r.* Without scientific attendance, they would practically always remain socially dangerous individuals. If we want to grow out of the present infantilism, experimentation on humans should be encouraged. Modern experimentation on animals is very humane, and suffering is eliminated. Criminals who are condemned to death should be given to science for experimenting. They would not suffer. Ultimately, they would probably die, but the benefits to the rest of mankind through scientific discoveries would be very important. Under present conditions, we 'take revenge on', 'punish'., mostly *sick* individuals, with seriously brutalizing semantic effects on the rest of us. There is not the slightest doubt that experimentation confined solely to animals, no matter how useful, will not solve many problems of Smith. Experimentation on humans is essential, and must be permitted. Most of the notorious criminals who go to the gallows appear at least infantile. How instructive it would be to make experiments on such individuals in respect to their thymus, . The list of experiments which science ought to make is very long, but material is lacking for such experimentation. Let me repeat that modern science can conduct its experiments without suffering to the individual, in spite of the fact that some of these experiments would be dangerous and might easily end in the painless death of the subject. The killing off of criminals (sick individuals) as a 'revenge' or 'punishment' or 'justice' is really too antiquated and too barbaric and *wasteful* for an enlightened society. If society wants to *eliminate* them, society can do it; but, at least, let us do it without such brutalizing morbidity, and with as great benefit to knowledge as possible.

The elimination of infantilism must be considered more than a personal issue with individuals; it becomes an *international semantic problem;* and such an international body as the League of Nations might originate a new era by starting a fundamental enquiry into this subject.

Infantilism in its national aspects is not equally distributed. Some countries are more infantile than others. In some countries even the university students show marked under-development for their age. Burrow reports that a questionnaire among students of a prominent university in the United States of America shows a surprisingly large percentage of onanism and homosexualism.

We should notice that not even all scientists are free from infantilism. Many of them are childlike in that they do not really care for science, or civilization, or society, but are *asocial* and merely like to play with their toys. As an excuse (rationalization of tendencies and 'emotions'), they usually profess 'science for science' sake', not realizing that a complete adult must become a *socialized* individual and cannot keep aloof from general human interests, and that science represents a *public*, time-binding activity and concern, not the private pleasure or benefit of some one person.

As we have already seen, a young child cannot be 'conscious of abstracting', but he can acquire it gradually with experience. Racial, ordered experience is called science. Every one of us has the tendencies, and, to some extent, the capacities, for developing science. The main aim of such racial, ordered experience is to save effort and unnecessary experiences, so that a child may start where the father leaves off (time-binding). The problems of consciousness of abstracting should be formulated by science and made available for semantic training. This would fulfill the main requirements of science, to save experience and effort, and to predict the future, to help in the mastery over external and internal 'nature', and so to produce semantic and physical adjustment.

If we teach and train the children in the consciousness of abstracting, we save them an enormous amount of the effort which would be necessary to acquire it eventually by themselves, and we also eliminate a great deal of unnecessary sufferings and disappointments. There is no danger of taking 'the joy out of life'; the opposite is true. With the consciousness of abstracting, the joy of living is considerably increased. We have no more 'frights', bewilderments, or similar undesirable semantic experiences. We grow up to full adulthood; and when the body is matured for the taking up of life and its responsibilities, we accomplish that, and find joy in it, as our 'mind' and 'emotions' have also matured. Such a consciousness of abstracting leads to an integrated, semantically balanced and adapted adult personality. Joys, pleasures, and 'emotions' are not abolished, as this cannot be done, given the structure of our nervous system and 'mental' health, but they are 'sublimated' to higher adult human semantic levels. Life becomes fuller, and the individual ceases to act as a nuisance and a danger to himself and others.

In the racial aspects, if the development of the individual became normal, we should grow beyond infantile organ erotic fixations and *el* languages and infantile systems in all fields. A [non-A]-system, in accordance with science 1933 ([non-E], [non-N] systems), would be the human link supplying scientific standards of evaluation to the affairs of Smith.

With the older infantilism and the practically general lack of full consciousness of abstracting, the fears, frights, painful 'emotional' shocks under which mankind lived were bound to have a marked, lasting, and sinister semantic and neurological effect upon the race. The race has never had an opportunity to develop in an adult way. What will be the results for the race of such a transformation it is impossible, at present, to foresee; but one thing is certain, that the results are bound to be very far-reaching.

19. CONCLUDING REMARKS

The world affairs have seemingly come to an impasse and probably, without the help of scientists, mathematicians, and psychiatrists included, we shall not be able to solve our urgent problems soon enough to prevent a complete collapse. Now those who are professionally engaged in human affairs, economists, sociologists, politicians, bankers, priests of every kind, teachers., 'mental' hygiene workers, and psychiatrists included, do not even suspect that material and methods of great general semantic value can be found in mathematics and the exact sciences. The drawing of their attention to this fact, no matter how clumsily done at first, will stimulate further researches, produce better formulations and understanding, and ultimately create conditions where sanity will be possible.

From a [non-A] point of view, a new era of human development seems possible, in which, by mere structural analysis and a linguistic revision, we will discover disregarded semantic mechanisms operating in all of us, which can be easily influenced and controlled; and we will discover, also, that at least a great deal of prevention can be accomplished.

It seems, also, that we will discover more about the dependence of 'human nature' on the structure of our languages, doctrines, institutions., and will conclude that for adjustment, stability., we must adjust these man-made and man-invented semantic and other conditions in comformity with that newly discovered 'human nature'. This, of course, would require a thorough scientific 1933, physico-mathematical, epistemological, structural, and semantic revision of all existing human interests, inclinations, institutions., to be made by those specialized in a 'science of man'. If such a revision is produced soon enough, it will, perhaps, help to adjust peacefully the standards of evaluation and prevent the repetition of bloody protests of unenlightened blind forces against *equally blind* forces of existing powers and reactions.

The forces of life, humanity, and time-binding are at odds; in modern slang, a 'show-down' is imminent; it *will happen*, and no one can prevent it. To a [non-A] understanding the only problem of importance is whether this 'show-down' will be scientific, enlightened, orderly, and peaceful, with minimum suffering; or whether it will take a blind, chaotic, silly, bloody, and wasteful turn with maximum suffering.

It becomes increasingly evident that we have come to a linguistic impasse, reflected in our historical, cultural, economic, social, doctrinal., impasses, all these issues being interconnected. The structural linguistic aspect is the most fundamental of them all, as it underlies the others and involves the *s.r*, or psycho-logical responses to words and other events in connection with meanings.

One of the benefits of building a system on undeniable negative premises is that many older and controversial problems become relatively simple and often uncontroversial, disclosing an important psycho-logical mechanism. Such formulations have often the appearance of the 'discovery of the obvious'; but it is known, in some quarters, that the discovery of the obvious is sometimes useful, not always easy, and often much delayed; as, for instance, the discovery of the equality of gravitational and inertial mass, which has lately revolutionized physics.

As words are *not* the things we are talking about, the only possible link between the objective world and the verbal world is structural. If the two structures are similar, then the empirical world becomes intelligible to us—we 'understand', we can adjust ourselves, . If we carry out verbal experiments and predict, these predictions are verified empirically. If the two structures are not similar, then our predictions are not verified —we do not 'know', we do not 'understand', the given problems are 'unintelligible' to us., we do not know what to do to adjust ourselves, .

Psychologically, in the first case we feel security, we are satisfied, hopeful.; in the second case, we feel insecure, a floating anxiety, fear, worry, disappointment, depression, hopelessness, and other harmful *s.r* appear. The considerations of structure thus disclose an unexpected and powerful semantic mechanism of individual and collective happiness, adjustment., but also of tragedies, supplying us with *physiological* means for a certain amount of desirable control, because relations and structure represent fundamental factors of all meanings and evaluations, and, therefore, of all *s.r.*

The present increasing world unrest is an excellent example of this. The structure of our old languages has shaped our *s.r* and suggested our doctrines, creeds., which build our institutions, customs, habits, and, finally, lead fatalistically to catastrophes like the World War. We have learned long ago, by repeated sad experience, that predictions concerning human affairs are not verified empirically. Our doctrines, institutions, and other disciplines are unable somehow to deal with this semantic situation, and hence the prevailing depression and pessimism.

We hear everywhere complaints of the stupidity or dishonesty of our rulers, as already defined, without realizing that although our rulers are admittedly very ignorant, and often dishonest, yet the most informed, gifted, and honest among them cannot predict or foresee happenings, if their arguments are performed in a language of a structure dissimilar to the world *and* to our nervous system. Under such conditions, calling names, even under provocations, is not constructive or helpful enough. Arguments in the languages of the old structure have led fatalistically to systems which are structurally 'un-natural' and so must collapse and impose unnecessary and artificial stress on our nervous system. The self-imposed conditions of life become more and more unbearable, resulting in the increase of 'mental' illness, prostitution, criminality, brutality, violence, suicides, and similar signs of maladjustment. It should never be forgotten that human endurance has limits. Human 'knowledge' shapes the human world, alters conditions, and other features of the environment—a factor which does not exist to any such extent in the animal world.

We often speak about the influence of heredity, but much less do we analyse what influence environment, and particularly the *verbal environment,* has upon us. Not only are all doctrines verbal, but the structure of an old language reflects the structural metaphysics of bygone generations, which affect the *s.r.* The vicious circle is complete. Primitive mythology shaped the structure of language. In it we have discussed and argued our institutions, systems., and so again the primitive structural assumptions or mythologies influenced them. It should not be forgotten that the affective interplay, interaction, interchange is ever present in human life, excepting, perhaps, in severe and comparatively rare (not in all countries) 'mental' ills. We can stop talking, we can stop reading or writing, and stop any 'intellectual', interplay and interaction between individuals, but we cannot stop or entirely abolish some *s.r.*

A structural linguistic readjustment will, it is true, result in making the majority of our old doctrines untenable, leading also to a fundamental scientific revision of new doctrines and systems, affecting all of them and our *s.r* in a constructive way. It is incorrect, for instance, to use the terms 'capitalism' as opposed to 'socialism', as these terms apply to different non-directly comparable aspects of the human problem. If we wish to use a term emphasizing the *symbolic character* of human relations, we can use the term 'capitalism', and then we can contrast directly individual, group, national, international., capitalisms. If we want to emphasize the psycho-logical aspects, we can speak of individualism versus socialism, . Obviously, in life the issues overlap, but the verbal implications remain, preventing clarity and inducing inappropriate *s.r* in any discussion.

In vernacular terms, there is at present a 'struggle' and 'competition' between two entirely different 'industrialisms' and two different 'commercialisms', based ultimately on two different forms of 'capitalism'. One is the 'individual capitalism', rapidly being transformed into 'group capitalism', in the main advanced theoretically to its limits in the United States of America and to a lesser extent in the rest of the civilized world, and 'social capitalism', proclaimed in the United Socialistic Soviet Republics. Both these extreme tendencies, connected also with semantic disturbances, are due to a verbal or doctrinal 'declaration of independence' of two, until lately, much isolated countries. The United States of America proclaimed the doctrine that man is 'free and independent', while, in fact, he is *not* free, but is inherently *interdependent*. The Soviets accepted uncritically an unrevised antiquated doctrine of the 'dictatorship of the proletarians'. In *practice*, this would mean the dictatorship of unenlightened masses, which, if left *actually to their creeds*, and deprived of the *brain-work* of scientists and leaders, would revert to primitive forms of animal life. Obviously, these two extreme creeds violate every typically *human* characteristic. We are interdependent, time-binders, and we are interdependent because we possess the higher nervous centres, which complexity animals do not possess. Without these higher centres, we could not be human at all; both countries seem to disregard this fact, as in both the brain-work is exploited, yet the brain-workers are not properly evaluated. The ignorant mob, with its historically and psycho-logically cultivated animalistic *s.r*, retards human progress and agreement. Leaders do not lead, but the majority play down to the mob psycho-logics, in fear of their heads or stomachs.

In both countries, the *s.r* are such that brain-work, although commercially exploited, is not properly evaluated, and is still persecuted here and there. For instance, in the United States of America, we witness court trials and resolutions against the work of Darwin, in spite of the fact that without some theory of evolution most of the natural sciences, medicine included, would be impossible. In Russia, we find decrees against researches in pure science, without which *modern* science is impossible. Both countries seemingly forget that all 'material' progress among humans is due uniquely to the *brain-work* of a few mostly underpaid and overworked workers, who exercise properly their higher nervous centres. With science getting hold of problems of *s.r* and sanity, our human relations and individual happiness will also become the subject matter of scientific enquiry. If international and *inter*dependent brain-workers produce discoveries and inventions, any one, even of the lowest development, can use or misuse their achievements, no matter what 'plan', or 'no-plan', is adopted. Both countries seem at present not to understand that a great development of mechanical means and the application of scientific achievements exclusively for animal comfort fail to lead to greater happiness or higher culture, and that, perhaps, indeed, they lead in just the opposite direction. Personally, I have no doubt that some day they will understand it; but an earlier understanding of this simple semantic fact would have saved, in the meantime, a great deal of suffering, bewilderment, and other semantic difficulties to a great number of people, if the rulers in both countries would be enlightened enough and could have foreseen it soon enough.

The future will witness a struggle between the individual and group capitalism, as exemplified in the United States of America, and the collective or social capitalism, as exemplified in the Soviet Republics. It does not require prophetic vision to foresee that some trends of history are foregone conclusions because of the structure of the human nervous system. As trusts or groups have replaced the theoretically 'individual' capitalism in the United States of America, so will the state capitalism replace the trusts, to be replaced in its turn by international capitalism.

We are not shocked by the international character of science. We are not '100 per cent patriotic' when it comes to the use in daily life of discoveries and inventions of other nations. Science is a semantic product of a *general human symbolic characteristic*; so, naturally, it must be general and, therefore, 'international'. But 'capitalism' is also a unique and *general* semantic product of symbolism; it is also a unique product of the human nervous system, dependent on mathematics, and, as such, by its inherent character, must become some day international. There is no reason why our *s.r* should be disturbed in one case more than in the other. The ultimate problem is not whether to 'abolish capitalism' or not, which will never happen in a symbolic class of life, but to transfer the control from private, socially irresponsible, uncontrolled, and mostly ignorant, leaders to more responsible, *professionally trained*, and socially controlled *public servants*, not bosses. If a country cannot produce honest, intelligent, and scientifically trained public men and leaders, that is, of course, very disastrous for its citizens; but this is not to be proclaimed as a rule, because it is an exception. Thus, in the Soviet Republics, graft is practically non-existent in the sense that it exists in the United States; but the mentality of the public men is practically at a similar standstill because of a deliberate minimizing of the value of brain-work. I wonder if it is realized at all, in either country, that *any* 'manual worker', no matter how lowly, is hired *exclusively* for his *human brain*, his *s.r,* and *not* primarily for his hands!

The only problem which the rest of mankind has to face is how this struggle will be managed and how long it will last, the outcome admitting of no doubt, as the ruthless elimination of individual capitalism by group capitalism (trusts) in the United States is an excellent example. In the Soviet Republics, they simply have gone further, but in a similar direction. Struggles mean suffering; and we should reconcile ourselves to that fact. If we want the minimum of suffering, we should stop the animalistic methods of contest. Human methods of solving problems depend on higher order abstractions, scientific investigations of structure and language, revision of our doctrines., resulting in peaceful adjustment of living facts, which are actualities whether we like it or not. If we want the maximum of suffering, let us proceed in the stupid, blind, animalistic and unscientific way of trial and error, as we are doing at present.

My aim is not to be a prophet, but to analyse different structural and linguistic semantic issues underlying all human activities, and so to produce material which may help mankind to *select* their lot *consciously*. What they will do is not my official concern, but it seems that both countries, which have so much in common, and which are bound to play an important role in the future of mankind, owing to their numbers, their areas, and their natural resources., will have to pay more attention to the so-called 'intellectual' issues, or, more simply, not disregard the difference between the reactions of infants and adults. Otherwise, very serious and disastrous cultural results for all of us will follow.

The problems of the world 1933 are acute and immediate, overloaded with confusion, bitterness, hopelessness, and other forms of semantic disturbances. Without some means—and, in this case, scientific and physiological means—to regulate our *s.r,* we shall not be able to solve our problems soon enough to avoid disasters. The similarity in structure of mathematics, and our nervous system, once pointed out and *applied*, gives us a unique means to regulate the *s.r,* without which it is practically impossible to analyse dispassionately and wisely the most pressing problems of immediate importance.

The present investigation shows that the old languages which, in structure, are *not* similar to the world and our nervous system, have automatically reflected their structure on our doctrines, creeds, and habits, *s.r,* and also on those man-made institutions which result from verbal arguments. These, in turn, shape further *s.r* and, as long as they last, control our destinies.

Four important issues could be shown in detail, but, for lack of space, I give only a suggestive sketch of them here.

1) In the [A]-system, all our existing older sub-systems, with all their benefits as well as shortcomings, follow as an [A] psycho-logical structural semantic necessity.

2) The tremendous handicap for any new and less deficient systems consists in the fact that such systems lack new constructive ∞-valued semantics, and are carried on the one side by linguistic two-valued arguments in the language of old *el* structure; yet they aspire 'emotionally' to something new and better, while the two cannot be reconciled.

3) An argument carried on in the old *el* and two-valued way, no matter how fundamentally true and eventually beneficial, can be easily defeated on verbal grounds if it follows the old structure of language. Our decisions are never well-grounded psycho-logically, and so can never command the respect or achieve the reliability of scientific reasoning. That is why we are groping— the only method possible under such conditions being the animalistic trial-and-error methods, swaying masses by inflammatory speeches because reason has nothing to offer, being tied up by the old verbal structure to the older consequences based on animalistic and fundamentally false-for-man assumptions.

4) In the old [A], *el*, two-valued system, agreement is theoretically impossible; so one of the main, and perhaps revolutionary, semantic departures from the old system is the fact that a *non-el* ∞-valued [non-A]-system, based on fundamental *negative* premises, leads to a theory of *universal agreement*, which is based on a structural revision of our languages, producing new and undisturbed *s.r*, which eliminate the copying of animals in our nervous reactions.

The problems of structure, language, and 'consciousness of abstracting' play a crucial semantic role. To be modern, one must accept modern metaphysics and a structurally revised modern language. As yet, these semantic problems have been *completely* disregarded as far as general education is concerned. This is probably due to the fact that in an infantile and commercial civilization we encourage engineering and applied sciences, medicine, biology., to increase private profits., and preserve or increase the ranks of buyers. But we do not encourage to an equal extent branches of science like mathematics, mathematical philosophy, linguistic, structural, and semantic researches., which would not directly increase profits or the numbers of customers, but which would, nevertheless, discover structural means for more happiness for all.

Accidentally—and this is recommended to the attention of economists—the classical law of 'supply and demand' is structurally and semantically an *animalistic law*, which in an adult human civilization must be reformulated. In fact, an adult human civilization cannot be produced at all if we preserve such fundamental animalistic 'laws'. In the animal world the numbers of individuals cannot increase beyond what the given conditions allow. The animals do not produce artificially.

Not so with our human world. We produce artificially because we are time-binders, and all of us stand on the shoulders of others and on the labours of the dead. We can over-populate this globe as we have done. Our numbers are not controlled by unaided nature, but can be increased considerably. In the animal world the numbers are regulated by the supply of food., and not by conditions imposed by the animals on that food supply. The animal law of 'supply and demand' is strict. In a human class of life, which does produce artificially, production should satisfy the wants of all, or their number should be controlled until the wants can be filled. The application of animalistic laws to ourselves makes conditions very complicated, and detrimental to most, if not all of us. It is also easily understood why it should be so. Ignorant and [A] handling of powerful symbols has proved to be dangerous when we do not realize the overwhelming semantic role and the importance of symbols in a symbolic class of life.

Another interesting application of the consciousness of abstracting is given in our attitude toward money, bonds, titles to property, . Money represents a symbol for all human time-binding characteristics. Animals do not have it. No doubt bees produce honey, but these products of the bees do not constitute wealth until man puts his hands on them. Money is not edible or habitable. It is worthless if the other fellow refuses to take it. The *m.o reality* behind the symbol is found in *human agreement*. The *value* behind the symbol is *doctrinal*. Fido does not discriminate between the different orders of abstractions. If we copy him, we worship the symbol alone. 'In gold we trust' becomes the motto, with all its identifications and destructive consequences. Smith should not identify the *m.o* reality behind the symbol with the symbol. It is amusing, when not tragic, to see how the so-called 'practical man' deals mostly with fictitious values, for which he is willing to live and die. When he has the upper hand and ignorantly plays with symbols, disregarding the *m.o* realities behind them, of course, he drives civilization to disasters. History is full of examples of this.

We see the utter folly of racing to accumulate symbols, worthless in themselves, while destroying the 'mental' and 'moral' values which are behind the symbols. For it is useless to 'own' a semantically unbalanced world. Such ownership is a fiction, no matter how stable it may look on paper. Commercialism, as a creed, is a folly of this type. Some day even economists, bankers, and merchants will understand that such 'impractical' works as this present one on structure, *s.r.*, lead to the revision of standards of evaluation and are directly helping the stabilization of an economic system. Meanwhile, in their ignorance, they do their best to keep the economic system unscientific, and, therefore, unbalanced. History shows clearly how the rulers have generally made life unbearable for the rest of mankind, and what bloody results have followed. Since the World War certain conditions are becoming increasingly more difficult, and the infantile and animalistic systems drive us fatalistically toward further catastrophes. Whether these disasters will occur, the unknown future shall decide; but out of this unknown, one fact remains a certainty; namely, that this will depend on whether or not science can take hold of human affairs; I hope it can, but the blind forces of identification are so strong and powerful that perhaps such hopes are premature. Perhaps a new race can accomplish it after this one is extinct, with the exception of a few remnants in museums.

To summarize, under present world conditions the role of governments is becoming more and more difficult and important. With all modern complexities it, is impossible for governmental men to be specialists in. every field of science, and therefore they must depend on professional experts *attached to the government*, not only in the fields of chemistry, engineering, physics, agriculture, etc., which they already utilize; but also in anthropology, neuro-psychiatry, general semantics, and related professions. Otherwise the governments will indefinitely play the role of the blind leading the blind. It is unreasonable to wait ten or twenty years to learn by bitter experience how short-sighted and incompetent our governments have been. Why not utilize some human intelligence, proper evaluation, etc., toward which extensional methods lead, and thereby have some *predictability*. *This is* definitely an imperative, immediate need.

We should not delude ourselves. Once the psychopathological misuses of neuro-semantic and neuro-linguistic mechanisms have been so successfully introduced, they will remain with us unless reconstructive and preventive governmental measures are undertaken by experts, at once.

The conditions of the world are such today that private scientific undertakings and even professional opinions of scientific societies, or international congresses, etc., are bound to be ineffective. Only governmental interest, backing, financing, etc., can organize and enforce a serious movement for sanity, the more so since scientists, physicians, educators, and other professionals do not have the necessary time, money, authority, or even initiative to carry forward concerted plans. We have learned this group wisdom by now in the case of smallpox vaccination, control of epidemics, etc., and I venture to suggest that only such group wisdom will be effective as far as the health of our nervous systems is concerned. In terms of money certainly it would be economical to spend for *preventive* and *permanent* measures an amount even less than the cost of a single aeroplane which is made today and shot down tomorrow.

It must be sadly admitted that even professionals, no matter how prominent they may be in their narrow specialties, as individuals or specialized groups are at present scientifically unequipped to deal with such large and complex problems as the passing from one system of orientation to another, because those whose duty it was to integrate methodologically the vast knowledge at hand, have failed. Such conditions can be remedied only by diversified methodological investigations, co-operation, and *concerted action of* specialists in different fields, which no private undertaking can organize effectively.

There can be little doubt that self-seeking politicians, to cover up their own tracks, will be against such scientific sanity guidance, but enlightened public opinion will sooner or later force the issues to the only possible intelligent solution.

The prevalent and constantly increasing general deterioration of human values is an unavoidable consequence of the crippling misuse of neuro-linguistic and neuro-semantic mechanisms. In general semantics we are concerned with the *sanity* of the race, including particularly methods of prevention; eliminating from home, elementary, and higher education inadequate aristotelian types of evaluation, which too often lead to the *un-sanity* of the race, and building up for the first time a positive theory of sanity, as a workable non-aristotelian *system*.

The task ahead is gigantic if we are to avoid more personal, national, and even international tragedies based on unpredictability, insecurity, fears, anxieties, etc., which are steadily disorganizing the functioning of the human nervous system. Only when we face these facts fearlessly and intelligently may we save for future civilizations whatever there is left to save, and build from the ruins of a dying epoch a new and saner society.

I seriously appeal to scientists, educators, medical men, especially psychopathologists, parents, and other forward-looking citizens to investigate and co-operate in urging the governments to carry out their duty to guide the people scientifically, as suggested here.

A non-aristotelian re-orientation is inevitable; the only problem today is when, and at what cost.

Appendix I

In an attempt to convey the magnitude of the task we are now confronting, I can do no better than to summarize roughly in the following tabulation some of the more outstanding points of difference between the aristotelian system as it shapes our lives today, and is lived by; and a scientific, non-aristotelian system, as it will, perhaps, guide our lives sometime in the future.

OLD ARISTOTELIAN ORIENTATIONS (CIRCA 350 B.C.)	NEW GENERAL SEMANTIC NON-ARISTOTELIAN ORIENTATIONS (1941 A.C)
1. Subject-predicate methods	Rational methods
2. Symmetrical relations, inadequate for proper evaluation	Asymmetrical relations, indispensable for proper evaluation
3. Static, 'objective', 'permanent', 'substance', 'solid matter', etc., orientations	Dynamic, ever-changing, etc., electronic process orientations
4. 'Properties' of 'substance', 'attributes', 'qualities' of 'matter', etc.	Relative invariance of function, dynamic structure, etc.
5. Two-valued, 'either-or', inflexible, dogmatic orientations	Infinite-valued flexibility, degree orientations
6. Static, finalistic 'allness'; finite number of characteristics attitudes	Dynamic non-allness; infinite number of characteristics attitudes
7. By definition 'absolute sameness in "all" respects' ('identity')	Empirical non-identity, a natural law as universal as gravitation
8. Two-valued 'certainty', etc.	Infinite-valued maximum probability
9. Static absolutism	Dynamic relativism
10. By definition 'absolute emptiness', 'absolute space', etc.	Empirical fullness of electro-magnetic, gravitational, etc, fields
11. By definition 'absolute time'	Empirical space-time
12. By definition 'absolute simultaneity'	Empirical relative simultaneity
13. Additive ('and'), linear	Functional, non-linear
14. (3+1)—dimensional 'space' and 'time'	4-dimensional space-time
15. Euclidean system	Non-euclidean systems
16. Newtonian system	Einsteinian or non-newtonian systems
17. 'Sense' data predominant	Inferential data as fundamental new factors
18. Macroscopic and microscopic levels	Sub-microscopic levels
19. Methods of magic (self-deception)	Elimination of self-deception
20. Fibers, neurons, etc., 'objective' orientations	Electro-colloidal process orientations
21. Eventual 'organism-as-a-whole', disregarding environmental factors	Organism-as-a-whole-in-environments, introducing new unavoidable factors
22. Elementalistic structure of language	
23. and orientations	Non-elementalistic structure of language and orientations
24. 'Emotion' and 'intellect', etc.	Semantic reactions
25. 'Body' and 'mind', etc.	Psychosomatic integration

26. Tendency to split 'personality'	Integrating 'personality'
27. Handicapping nervous integration	Producing automatically thalamo-cortical integration
28. Intensional structure of language and orientations, perpetuating:	Extensional structure of language and orientations, producing:
29. Identifications in value: a) of electronic, electro-colloidal, etc., stages of processes with the silent, non-verbal, 'objective' levels b) of individuals, situations, etc. c) of orders of abstractions	Consciousness of abstracting \\ | > Extensional devices | /
30. Pathologically reversed order of evaluation	Natural order of evaluation
31. Conducive to neuro-semantic tension	Producing neuro-semantic relaxation
32. Injurious psychosomatic effects	Beneficial psychosomatic effects
33. Influencing toward un-sanity	Influencing toward sanity
34. 'Action at a distance', metaphysical false-to-fact orientations	'Action by contact', neuro-physiological scientific orientations
35. Two-valued causality, and so consequent 'final causation'	Infinite-valued causality, where the 'final causation' hypothesis is not needed
36. Mathematics derived from 'logic', with resulting verbal paradoxes	'Logic' derived from mathematics, eliminating verbal paradoxes
37. Avoiding empirical paradoxes	Facing empirical paradoxes
38. Adjusting empirical facts to verbal patterns	Adjust verbal patterns to empirical facts
39. Primitive static 'science' (religions)	Modern dynamic 'religions' (science)
40. Anthropomorphic	Non-anthropomorphic
41. Non-similarity of structure between language and facts	Similarity of structure between language and facts
42. Improper evaluations, resulting in:	Proper evaluations, tested by:
43. Impaired predictability	Maximum predictability
44. Disregarded	Undefined terms
45. Disregarded	Self-reflexiveness of language
46. Disregarded	Multiordinal mechanisms and terms
47. Disregarded	Over/under defined character of terms
48. Disregarded	Inferential terms as terms
49. Disregarded	Neuro-linguistic environments considered as environment
50. Disregarded	Neuro-semantic environments considered as environment
51. Disregarded	Decisive, automatic effect of the structure of language on types of evaluation, and so neuro-semantic reactions
52. Elementalistic, verbal, intensional, 'meaning', or still worse, 'meaning of meaning'	Non-elementalistic, extensional, by fact evaluations
53. Antiquated	Modern, 1941

Appendix II

Table of Contents: Fifth Edition of *Science and Sanity*
A Note on Errata
Preface to the Fifth Edition by Robert P. Pula

Preface to the Fourth Edition by Russell Meyers, M.D.
Bibliographical Note, 1958

Preface to the Third Edition 1948

Introduction to the Second Edition 1941

A.   Recent Developments and the Founding of the Institute of General Semantics
B.   Some Difficulties to be Surmounted
            1. Attitudes of 'Philosophers', etc.
            2. Perplexities in theories of 'meaning'
            3. Inadequacy of forms of representation and their structural revision
            4. Identifications and mis-evaluations
            5. Methods of the magician
C.   Revolutions and Evolutions
D.   A Non-aristotelian Revision (Tabulated)
E.   New Factors: The Havoc they Played with our Generalizations
F.   Non-aristotelian Methods
            1. Neurological mechanisms of extensionalization
            2. Neuro-semantic relaxation
            3. Extensional devices and some applications
            4. Implications of the structure of language
G.   Over/Under-defined Terms
H.   The Passing of the Old Aristotelian Epoch
            1. 'Magnot Line Mentalities'
            2. Wars of and on nerves
            3. Hitler and psycho-logical factors in his life
            4. Education for intelligence and democracy
I.   Constructive Suggestions
J.   Conclusion

Supplementary Bibliography
Preface to the First Edition 1933

Book I
A General Survey of Non-aristotelian Factors
Part I
Preliminaries

I.   Aims, Means, and Consequences of a Non-aristotelian Revision
II.   Terminolgy and Meanings On Semantic Reactions
            A. On the Un-speakable Objective Level
            B. On "Copying" Nervous Reactions
III.   Introduction

SUPPLEMENT
COLLOIDAL BEHAVIOUR

In fact, to-day colloids may be regarded as an important, perhaps the most important connecting link between the organic and the inorganic world. — WOLFGANG PAULI

In our researches, let us follow the natural order and give a brief structural account of what we know, empirically, about the medium in which life is found; namely, about the colloids. The following few elementary particulars show the empirical importance of structure, and so are fundamental in the present work.

At present, physicians are usually too innocent of psychiatry, and psychiatrists, although they often complain about this innocence of their colleagues, seldom, if ever, themselves pay any attention to the colloidal structure of life; and their arguments about the 'body-mind' problem are still scientifically incomplete and unconvincing, though the 'bodymind' problem has been present with us for thousands of years. It is a very important semantic problem, and, as yet, not solved scientifically, although there is a simple solution of it to be found in the colloidal structure of life.

The reader should not ascribe any uniqueness of the 'cause-effect' character to the statements which follow, as they may not be true when generalized. Colloidal science is young and little known. Science has accumulated a maze of facts, but we do not have, as yet, a general theory of colloidal behaviour. Statements, therefore, should not be unduly generalized.

We shall only indicate a few structural and relational connections important for our purpose.

When we take a piece of some material and subdivide it into smaller pieces, we cannot carry on this process indefinitely. At some stage of this process the bits become so small that they cannot be seen with the most powerful microscope. At a further stage, we should reach a limit of the subdivision that the particles can undergo without losing their chemical character. Such a limit is called the molecule. [This statement is only approximate, because there is evidence that chemical characteristics change as the molecule is approached. The smallest particle visible in the microscope is still about one thousand times larger than the largest molecule. So we see that between the molecule and the smallest visible particle there is a wide range of sizes. Findlay calls these the 'twilight zone of matter'; and it was Oswald, I believe, who called it the 'world of neglected dimensions'.

This 'world of neglected dimensions' is of particular interest to us, because in this range of subdivision or smallness we find very peculiar forms of behaviour-life included-which are called 'colloidal behaviour'.

The term 'colloid' was proposed in 1861 by Thomas Graham to describe the distinction between the behaviour of those materials which readily crystallize and diffuse through animal membranes and those which form 'amorphous' or gelatinous masses and do not diffuse readily or at all through animal membranes. Graham called the first class 'crystalloids' and the second 'colloids', from the Greek word for glue.

In the beginning colloids were regarded as special 'substances', but it was found that this point of view was not correct. For instance, NaCl may behave in solution either as a crystalloid or as a colloid; so we began to speak about the colloidal state. Of late, even this term became unsatisfactory and is often supplanted by the term 'colloidal behaviour'.

In general, a colloid may be described as a 'system' consisting of two or more 'phases'. The commonest represent emulsions or suspensions of fine particles in a gaseous, liquid, or other medium, the size of the particles grading from those barely visible microscopically to those of molecular dimensions. These particles may be either homogeneous solids, or liquids, or solutions themselves of a small percentage of the medium in an otherwise homogeneous complex. Such solutions have one characteristic in common ; namely, that the suspended materials may remain almost indefinitely in suspension, because the tendency to settle, due to gravity, is counteracted by some other factor tending to keep the particles suspended. In the main, colloidal behaviour is not dependent upon the physical state or chemistry of the finely subdivided materials or of the medium. We find colloidal behaviour exhibited not only by colloidal suspensions and emulsions where solid particles or liquid droplets are in a liquid medium, but also when solid particles are dispersed in gaseous medium (smokes), or liquid droplets in gaseous media (mists),.Materials which exhibit this special colloidal behaviour are always in a very fine state of subdivision, so that the ratio of surface exposed to volume of material is very large. A sphere containing only 10 cubic centimetres, if composed of fine particles 0.00000025 cm. in diameter, would have a total area of all the surfaces of the particles nearly equal to half an acre.' It is easy to understand that under such structural conditions the surface forces become important and play a prominent role in colloidal behaviour.

The smaller the colloidal particles, the closer we come to molecular and atomic sizes. Since we know atoms represent electrical structures, we should not be surprised to find that, in colloids, surface energies and electrical charges become of fundamental importance, as by necessity all surfaces are made up of electrical charges. The surface energies operating in finely grained and dispersed systems are large, and in their tendency for a minimum, every two particles or drops tend to become one; because, while the mass is not altered by this change, the surface of one larger particle or drop is less than the surface of two smaller ones--an elementary geometrical fact. Electrical charges have the well-known characteristic that like repels like and attracts the unlike. In colloids, the effect of these factors is of a fundamental, yet opposite, character. The surface energies tend to unite the particles, to coagulate, flocculate or precipitate them. In the meanwhile the electrical charges tend to preserve the state of suspension by repelling the particles from each other. On the predominance or intensity of one or other of these factors, the instability or the stability of a suspension depends.

In general, if 'time' limits are not taken into consideration, colloids are unstable complexes, in which continuous transformation takes place, which is induced by light, heat, electric fields, electronic discharges, and other forms of energy. These transformations result in a great variation of the characteristics of the system. The dispersed phase alters its characteristics and the system begins to coagulate, reaching a stable state when the coagulation is complete. This process of transformation of the characteristics of the system which define the colloid, and which ends in coagulation, is called the 'ageing' of the colloid. With the coagulation complete, the system loses its colloidal behaviour-it is 'dead'. Both of these terms apply to inorganic as well as to organic systems.

Some of the coagulating processes are partial and reversible, and take the form of change in viscosity; some are not. Some are slow; some extremely rapid, particularly when produced by external agencies which alter the colloidal equilibrium.

From what has been said already, it is obvious that colloids, particularly in organisms, are extremely sensitive and complex structures with enormous possibilities as to degree of stability, reversibility., and allow a wide range of variation of behaviour. When we speak of 'chemistry', we are concerned with a science which deals with certain materials which preserve or alter certain of their characteristics. In 'physics', we go beyond the obvious characteristics and try to discover the structure underlying these characteristics. Modern researches show dearly that atoms have a very complex structure and that the macroscopic characteristics are directly connected with sub-microscopic structure. If we can alter this structure, we usually can alter also the chemical or other characteristics. As the processes in colloids are largely structural and physical, anything which tends to have a structural effect usually also disturbs the colloidal equilibrium, and then different macroscopic effects appear. As these changes occur as series of interrelated events, the best way is to consider colloidal behaviour as a physico-electro-chemical occurrence. But once the word 'physical' enters, structural implications are involved. This explains also why all known forms of radiant energy, being structures, can affect or alter colloidal structures, and so have marked effect on colloids.

As all life is found in the colloidal form and has many characteristics found also in inorganic colloids, it appears that colloids supply us with the most important known link between the inorganic and the organic. This fact also suggests entirely new fields for the study of theliving cells and of the optimum conditions for their development, sanity included.

Many writers are not agreed as to the use of the terms 'film', 'membrane', and the like. Empirically discovered structure shows clearly, however, that we deal with surfaces and surface energies and that a 'surface tension film' behaves as a membrane. In the present work, we accept the obvious fact that organized systems are film-partitioned systems.

One of the most baffling problems has been the peculiar periodicity or rhythmicity which we find in life. Lately, Lillie and others have shown that this rhythmicity could not be explained by purely physical nor purely chemical means, but that it is satisfactorily explained when treated as a physico-electro-chemical structural occurrence. The famous experiments of Lillie, who used an iron wire immersed in nitric acid and reproduced, experimentally, a beautiful periodicity resembling closely some of the activities of protoplasm and the nervous system, show conclusively that both the living and the non-living systems depend for their rhythmic behaviour on the chemically alterable film, which divides the electrically conducting phases. In the iron wire and nitric acid experiment, the metal and the acid represent the two phases, and between the two there is found a thin film of oxide. In protoplasmic structures, such as a nerve fibre, the internal protoplasm and the surrounding medium are the two phases, separated by a surface film of modified plasm membrane. In both systems,. the electromotive characteristics of the surfaces are determined by the character of the film.

That living organisms are film-bounded and partitioned systems accounts also for irritability. It appears that irritability manifests itself as sensitiveness to electrical currents. These currents seem to depend on polarizability or resistance to the passage of ions, owing to the presence of semi-permeable boundary films or surfaces enclosing or partitioning the system. It is obvious that we are here dealing with complex structures which are intimately connected with the characteristics of life. Living protoplasm is electrically sensitive only as long as its structure is intact. With death, semi-permeability and polarizability are lost, together with electrical sensitivity.

One of the baffling peculiarities of organisms is the rapidity with which the chemical and metabolic processes spread. Indeed, it is impossible to explain this by the transportation of material. All evidence shows that electrical and, perhaps, other energy factors play an important role; and that this activity again depends on the presence of surfaces of protoplasmic structures with electrode like characteristics which form circuits.

The great importance of the electrical charges of the colloidal particles arises out of the fact that they prevent particles from coalescing; and when these charges are neutralized, the particles tend to form larger aggregates and settle out of the solution. Because of these charges, when an electrical current is sent through a colloidal solution, the differently charged particles wander to one or the other electrode. This process is called cataphoresis. There is an important difference in behaviour in inorganic and organic colloids under the influence of electrical currents, and this is due to the difference in structure. In inorganic colloids, an electrical current does not coagulate the whole, but only that portion of it in the immediate vicinity of the electrodes. Not so in living protoplasm. Even a weak current usually coagulates the entire protoplasm, because the inter-cellular films probably play the role of electrodes and so the entire protoplasm structurally represents the 'immediate vicinity' of the electrodes.' Similarly, structure also accounts for the extremely rapid spread of some effects upon the whole of the organism.

Electrical phenomena in living tissue are mainly of two more or less distinct characters. The first include electromotive energy which produces electrical currents in nerve tissue, the membrane potentials,. The second are called, by Freundlich, electrokinetic, and include cataphoresis, agglutination,. There is much evidence that the mechanical work of the muscles, the secretory action of the glands, and the electrical work of the nerve cells are closely connected with the colloidal structure of these tissues. This would explain why any factor (semantic reactions included) capable of altering the colloidal structure of the living protoplasm must have a marked effect on the behaviour and welfare of the organism.

Experiments show that there are four main factors which are able to disturb the colloidal equilibrium : (1) Physical, as, for instance, X-rays, radium, light, ultra-violet rays, cathode rays.; (2) Mechanical, such as friction, puncture.; (3) Chemical, such as tar, paraffin, arsenic.; and, finally, (4) Biological, such as microbes, parasites, spermatozoa,. In man, another (fifth) potent factor; namely, the semantic reactions,enters, but about this factor, I shall speak later.

For our purpose, the effects produced by the physical factors, because obviously structural, are of main interest, and we shall, therefore, summarize some of the experimental structural results. Electrical currents of different strength and duration, as well as acids of different concentration, or addition of metallic salts, which produce marked acidity, usually coagulate the protoplasm, these effects being structurally interrelated. Slow coagulation involves changes in viscosity, all of which, under certain conditions, may be reversible. When cells are active, their fluidity often changes in a sharp and rapid manner.

Fat solvents are called surface-active materials; when diluted, they decrease protoplasmic viscosity; but more concentrated solutions produce increased viscosity or coagulation." The anaesthetics, which always are fat solvents and surface-active materials, are very instructive in their action for our purpose, as they affect very diversified types of protoplasm similarly, this similarity of action being due to the similarity of colloidal structure. Thus, ether of equal concentration will make a man unconscious, will prevent the movement of a fish and the wriggling of a worm, or stop the activity of a plant cell, without permanently injuring the cells." In fact, the action of all drugs is based on their effect upon the colloidal equilibrium, without which action a drug would not be effective. It is well known that various acids or alkalis always change the electrical resistance of the protoplasm.'

The working of the organism involves mostly a structural and very important 'vicious circle', which makes the character of colloidal changes non-additive. If, for instance, the heart, for any reason, slows down the circulation, this produces an accumulation of carbonic acid in the blood, which again increases the viscosity of the blood and so throws more work on the already weakened heart. Under such structural conditions, the results may accumulate very rapidly, even at a rate which can be expressed as an exponential function of higher degree.

Different regions of the organism have different charges ; but, in the main, an injured, or excited, or cooler part is electro-negative (which is connected with acid formation), and the electro-positive particles rush to those parts and supply the material for whatever physiological need theree may be.'

The effects of different forms of radiant energy on colloids and protoplasm are being extensively studied, and the results are very startling. The different forms of radiant energy differ in wave-length, frequency.,—that is to say, generally in structure,— and, as such, may produce structural effects on colloids and organisms, which effects may appear on the gross macroscopic level in many different forms.

Electrical currents, for instance, retard reversibly the growth of roots, may activate some eggs into larval stages without fertilization., which makes it possible to understand why, in some cases, a mere puncturing of the egg may disturb the equilibrium and produce the effects of fertilization.

The X-, or Röntgen-rays have been shown to accelerate 150 times the process of mutation. Muller, in his experiments with several thousand cultures of the fruit fly, has established the above ratio of induced mutations, which become hereditary." 'Cosmic rays' in the form of radiation from the earth, in tunnels, for instance, show similar results, except that mutation occurs only twice as often as under the usual laboratory conditions. Under the influence of X-rays, mice change their colour of hair; gray mice become white, and white ones darker. Sometimes further additional bodily changes appear; as, for instance, one or no kidneys, abnormal eyes or legs, occur more often than under ordinary conditions. Some animals lose their power of reproduction, although the body is not obviously changed. Plants respond also to the X-ray treatment. They grow faster, flower more, and produce new forms more readily. In humans the effect of X-ray irradiation has often proven disastrous to the health of experimenters. There are even data that the irradiation of pregnant mothers may result in deformation of the head and limbs of the unborn child and, in one-third of the cases, feeble-mindedness of the children has resulted.

Ultra-violet rays also show a marked effect. In some instances, they slow down or stop the streaming of protoplasm, because of increased viscosity or coagulation; plants grow slowly or rapidly; certain valuable ingredients in plants are increased; certain animals, as, for instance, small crustacea or bacteria are killed; eggs of Nereis (a kind of sea worm), which usually have 28 chromosomes, after irradiation have 70; certain bone malformations in children are cured; the toxin in the blood serum of pernicious anaemia patients is destroyed,. In this respect, we should notice again that ultra-violet irradiation produces curative effects like those of cod-liver oil, which shows that the effect of both factors is ultimately colloidal and structural.

Extensive experimentation with cathode rays is very recent, but already we have a most astonishing arrayof structural facts. Moist air is converted into nitric acid, synthetic rubber is produced rapidly, the milk from rubber trees is made solid and insoluble without the use of sulphur, liquid forms of bakelite are solidified without heating, linseed oil becomes dry to the touch in three hours and hard in six hours, certain materials, like cholesterol, yeast, starch, cottonseed oil, after exposure for thirty seconds, heal rickets, and similar unexpected results. What are usually called 'vitamins' do not only represent 'special substances', but become structurally active factors ; and this is why ultra-violet rays may produce results like those of some 'substance'. It seems that in 'vita-mins' the surface activities are important; the parallelism shown by von Kahn between the surface activities of different materials and the Funk table of vitamin content is quite suggestive. Some data seem to show that, in some instances, surface-active materials, such as coffee or alcohol, produce beneficial surface activities similar to the 'vitamins'.'

The above short list gives only an approximate picture of the overwhelming importance of the roles which structure in general, and colloids in particular, play in our lives. We see about us many human types. Some are delicate, some are heavy-set, some flabby, some puffy, all of which indicates a difference in their colloidal structure. Paired with these physical colloidal states are also nervous, 'mental', and other characteristics, which vary from weak and nervous to the extreme limitation of nervous activities, as in idiocy, which is a negation of activity. It is curious that in all illnesses, whether 'physical' or 'mental', the symptoms are very few, and fundamentally of a standard type. In 'physical' illness we find the following common characteristics : fever, chills, headaches, convulsions, vomiting, diarrhoea,. In 'mental' ills, identifications, illusions, delusions, and hallucinations-in general, the reversed pathological order-are found. It is not difficult to understand the reason. Because of the general colloidal background of life, different disturbances of colloidal equilibrium should produce similar symptoms. In fact, many of these symptoms have been reproduced experimentally by injecting inert precipitates incapable of chemical reactions, which have induced artificial colloidal disturbances. Thus, if the serum from an epileptic patient is injected into a guinea pig, it results in an attack of convulsions, often ending in death. But, if the guinea pig is previously made immune by an injection of some colloid which accustoms the nerveendings to the colloidal flocculation, then, for a few hours following, we can, with impunity, introduce into the circulation otherwise fatal doses of epileptic serum. Epileptic serum can also be made immune by filtration, or by strong centrifugation, or by long standing, which frees it from colloidal precipitates.

Death through blood transfusion or the injection of any colloid-into the circulation has also, in the main, similar symptoms, regardless of the chemical character of the colloid, indicating once more the importance and fundamental character of structure.

That illnesses are somehow connected with colloidal disturbances (note the wording of this statement) becomes quite obvious when we consider catarrhal diseases, inflammations, swellings, tumours, cancer, blood thrombi., which involve colloidal injuries, resulting in extreme cases in complete coagulation or fluidification, the variation between 'gel' and 'sol' appearing in a most diversified manner."' Other illnesses are connected with precipitation or deposits of various materials. Gout, for instance, results from a morbid deposit of uric acid, and different concretions, such as the 'stones', are very often found in different fluids of the organism. We have, thus, concretions in the intestines, the bile, the urine, the pancreas, the salivary glands; lime deposits in old softened tissues, 'rice bodies' in the joints, 'brain sand',."

In bacterial diseases, the micro-organisms rapidly produce acids and bases which tend to destroythe colloidal equilibrium. Lately, it has been found that even tuberculosis is more than a mere chapter in bacteriology. All the main tubercular symptoms can be reproduced, experimentally, by means of colloidal disturbances without the intervention of a single bacterium.This would explain also why, in some instances, psychotherapy is effective in diseases with tubercular symptoms.

By structural necessity, every expression of cellular activity involves some sort of colloidal behaviour; and any factor disturbing the colloidal structure must be disturbing to the welfare of the organism. Vice versa, a factor which is beneficial to the organism must reach and affect the colloids.

After this brief account of the structural peculiarities of the domain in which life is found, we can understand the baffling 'body-mind' problem. We do not yet know as many details as we could wish, but these will accumulate the moment a general solution is clearly formulated. It is a well-established experimental fact that all nervous and 'mental' activities are connected with, or actually generate, electrical currents, which of late are scrupulously studied by the aid of an instrument called the psychogalvanometer.It is not suggested that electrical currents are the only ones which are involved. There may be many different formsof radiant energy produced or effective, which we have not yet the instruments to record. Experiments suggest such a possibility. Thus, for instance, the apex of a certain rapidly growing vegetable or animal tissue emits some sort of invisible radiation which stimulates the growth of living tissue with which it is not in contact. The tip of a turnip or onion root, if placed at right angles to another root, at a distance of a quarter of an inch, so stimulates the growth of the latter that the increase of the number of cells, on its side nearest the point of stimulation, is as high as seventy per cent. These radiations accelerate the growth of some bacteria. Other examples could be given.

A classical example of the effect left on protoplasm by energetic factors is given by Bovie. Yet, we have not assumed that the protoplasm of plants also shows lasting structural and functional results of stimulation, some sortof 'learning' or 'habit-formation' characteristics. But such is the case; and further experimentation along these lines will help greatly to under stand the mechanism of 'mental' processes in ourselves.

If we take the seed of a plant, for instance, of a squash, and keep it in a moist tropism chamber in the dark,, it will grow a root. When theroot is about one inch long, we begin our experiment. Originally, under the influence of gravitation, the root grows vertically downwards (A). If we rotate the tropism chamber 90° so that the root is horizontal (B), the root will soon bend down- wards under the influence of positive geotropism. But the bending does not occur at once. There is a latent period-in the case of the squash seed, about ten minutes-after which pause the root is bent

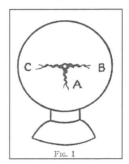

Fig. 1

downwards. When we have determined this latent period for a given seedling, we then rotate the chamber to the positions (B), (C), (A), (B), (C)., just within the 'time' limit before the bending. would occur. We repeat such procedure sev- eral times. When we set the root again in its vertical downward position (A), we notice that the root, without any more changes of position, will wag backwards and forward with the period as was used in the experiment. This unexpected behaviour will last for several days. It shows that the alternating stimulus of gravitation, as applied to the root, has produced some structural changes in the protoplasm which persist for a comparatively long period after the stimulus has ceased to act. It becomes obvious that teachability and the structural tendency for forming engrams is a general characteristic of protoplasm.

All the examples given above show clearly that structure in general, and of colloids in particular, gives us a satisfactory basis for the understanding of the equivalence between occurrences which belonged formerly to 'chemistry' and those classified as 'physical', and ultimately between these and those we call 'mental'. Structure, and structure alone, gives not only the unique content of what we call 'knowledge', but also the bridge between the different classes of occurrences—a fact which, as yet, has not been fully understood.

To sum up: It is known that colloidal behaviour is exhibited by materials of very fine subdivision, the 'world of neglected dimensions', which involves surface activities and electrical characters of manifold and complex structure, and therefore the flexibility of gross macroscopic characteristics. It is well known that all life-processes, 'feelings', 'emotions', 'thought', semantic reactions, and so forth, involve at least electrical currents. As electrical currents and other forms of energy are able to affect the colloidal structure on which our physical characteristics depend, obviously 'feelings', 'emotions', 'thought'.; in general, s.r, which are connected with manifestations of energy, will also have some effect on our bodies, and vice versa. Colloidal structure supplies us with an extremely flexible mechanism with endless possibilities.

When we analyse the known empirical facts from a structural point of view, we find not only the equivalence which was mentioned before, but we must, also, legitimately consider the so-called 'mental', 'emotional', and other semantic and nervous occurrences in connection with manifestations of energy which have a powerful influence on the colloidal behaviour, and so ultimately on the behaviour of our organisms as-a-whole. Under such environmental conditions, we must take into account all energies which have-been discovered, semantic reactions not excluded, as all such energies have structural effect. As language is one of the expressions of one of these energies, we ought to find it quite natural that the structure of language finds its reflection in the structure of the environmental conditions which are dependent on it.

Until lately, the disregard of colloidal science and of structure in general has greatly retarded advance in biology, psychiatry, and other sciences. Biology, for instance, has mostly studied 'life' where none existed; namely, in death. If we study corpses, we study death, not life, and life is a function of living cells. The living cell is semi-fluid, and all the forces which act in colloidal solutions and constitute colloidal behaviour are acting because they can act, while a dead cell is coagulated and so a different set of energies is operating there .

Should we wonder that life, being a form of colloidal behaviour on microscopic and sub-microscopic levels, conditioned by little colloidal 'wholes', and structures separated from their environment by surfaces, preserves a similar character on macroscopic levels? We should, instead, be surprised if this did not turn out to be the case.

INDEX

A. See Aristotelian system.

[non-A] See Non-Aristotelian system.

Abbreviation of terms, xx

Abstracting, ii, iii, vi-x, xiii, 8, 18, 20-21, 22 ff., 29, 41, 50-51, 56, 66, 69, 80, 83, 87-89, 91, 96, 99, 101, 105, 108, 110-113, 114-153, 157, 161-169, 173-174, 179-180, 183, 185-186

    accepted as fundamental, 51, 119-120

    consciousness of, ii, iii, vi, ix, xiii, 8, 18, 20-21, 41, 66, 69, 89, 96, 99, 101, 105, 110-113 128-135, 136-144, 147-152, 157, 161-169, 173-174, 179-180, 183, 186

        functional definition, 138

        in mathematics, 89, 104-105

        in primitives, 169

    functional definition, 136-138

    levels of. See Abstractions, orders of.

    limits of, in animals, 82, 110, 127-128, 136, 151

    model of. See Structural Differential.

    orders of. See Abstractions, orders of.

    process of, vi-x, 50-51, 88, 91, 114-153, 163

Abstractions, 'absolute,' order of, vi

    by children, 173

    by different 'senses,' 82, 91

    confusion of orders of, 38, 51-53, 59-60, 70, 81, 89-90, 99-100, 111, 114-116, 134, 139 142, 152, 156

    disregarding orders of, 151

    higher order, 23, 53, 57, 59, 70, 76, 80, 86-88, 91-92, 97-104, 108-109, 112-113, 118-121, 126-129, 134, 139, 144, 145-153, 155, 157, 160, 178, 186

    levels of. See orders of.

    macroscopic, 124, 129

    multiordinality of, 56

    orders of, vi, viii, 2, 11, 20, 42, 51-52, 55-56, 60, 69-72, 76, 87-92, 96, 100-101, 105, 107-108, 111-114, 118, 123, 124-135, 138, 141-153, 156-159, 163, 168, 180, 183, 186

'Acquaintance' introduced as a non-el term, 116-117

Activation of nerve currents, 47-49, 53, 59, 63, 90, 98-103, 142, 158, 163

Additivity, 72, 89

Adulthood of humanity, 2, 163

Affective field and disturbances, 102, 104, 141

    origins of 'mental' ills, 43-44

    tension, 157

Aggregates, 92-95, 185, 190. See also Theory, set.

Agriculture, 180

'Allness', viii, 51, 117, 120, 133, 138, 156, 161-164, 169, 182, 188

Analogies. See also Illustrative Examples.

    emery sand lubricant, 3-4

    fan blades and submicroscopic levels, 121-122, 162

    identification-disease, 3-4

    lepers and 'normal men', 111

CPSIA information can be obtained
at www.ICGtesting.com
Printed in the USA
LVHW060226291221
707217LV00003B/49